Advances in Experimental Medicine and Biology

Volume 807

For further volumes:
http://www.springer.com/series/5584

Advances in Experimental Medicine and Biology

Rameshwar Adhikari · Santosh Thapa
Editors

Infectious Diseases and Nanomedicine I

First International Conference (ICIDN-2012), Dec. 15–18, 2012, Kathmandu, Nepal

Springer

Editors
Rameshwar Adhikari
Central Department of Chemistry
Tribhuvan University
Kathmandu
Nepal

Santosh Thapa
Department of Microbiology
Tribhuvan University
Kathmandu
Nepal

ISSN 0065-2598 ISSN 2214-8019 (electronic)
ISBN 978-81-322-1776-3 ISBN 978-81-322-1777-0 (eBook)
DOI 10.1007/978-81-322-1777-0
Springer New Delhi Heidelberg New York Dordrecht London

Library of Congress Control Number: 2014930226

Printed on acid-free paper

Springer is part of Springer Science+Business Media (www.springer.com)

Preface

This Special Issue of *Advances in Experimental Medicine and Biology* is devoted to the First International Conference on Infectious Diseases and Nanomedicine (ICIDN) which was innovatively conceptualized by the Nepalese Association of Medical Microbiology (NAMM) and the Nepal Polymer Institute (NPI) in association with Kathmandu University, Dhulikhel, Nepal to promote interdisciplinary scientific study for systemic understanding of connections between major human diseases and their treatment regime by applying the tools and techniques of nanotechnology. The conference organized in Park Village Hotel and Resort, Budhanilkantha, Kathmandu from December 15 to 18, 2012, was accompanied by a one-day Short Course on *Recent Trends in Nanomedicine and Infectious Diseases* and had the motto: *Interdisciplinary Collaborative Education for Research & Innovation in Biomedical Sciences*. The ICIDN-2012 was attended by approximately 180 scientists from 21 countries including dozens of eminent scientists from throughout the globe.

One of the aims of this congress was to get researchers from the fields of nanomedicine (including physics, chemistry, and engineering) and various fields of infectious diseases (including pathology, microbiology, and clinical sciences) together at the same congress to explore common ground. The conference focused on themes on nanomedicine with special attention to anticancer therapy, antibiotic resistance, and virology.

The Inauguration Ceremony was attended by Rt. Hon'ble President of Nepal, Dr. Ram Baran Yadav as Chief Guest. In his inaugural address, President Yadav stressed on the role of young scientists and students in developing the nation through science, technology, and innovations. He praised the efforts of the ICIDN organizers to arrange such a global scientific meeting of enormous scientific and practical significance in Nepal. The inauguration ceremony was also addressed by Prof. Geoffrey L. Smith (President of the International Union of Microbiological Societies (IUMS); Head of the Department of Pathology, University of Cambridge, England), Prof. Jamboor K. Vishwanatha (Dean, Graduate School of Biomedical Sciences, University of North Texas Health Science Center, USA), Prof. Surendra Raj Kafle (Vice Chancellor of Nepal Academy of Science and Technology (NAST)), Prof. Ishwar Chandra Dutta (Chairman of Tribhuvan University Service

Commission), and Dr. Rajendra Koju (CEO, School of Medical Sciences, Kathmandu University, the representative of Vice Chancellor of the University). The congress was opened formally through the video lecture on *Helicobacter pylori* by Nobel Laureate Prof. Barry J. Marshall from The University of Western Australia.

The conference served as a meeting point for young researchers and students with their peers and well-known scientists for making scientific contacts and exchanging their experiences. The enthusiastic participation of numerous young scientists and students from all over the world led to vivid scientific discussions during the breaks, poster sessions, as well as after the sessions. Ms. Pramila Adhikari, Ms. Manisha Pradhan, and Mr. Eak Dev Khanal were bestowed best Poster Presentation Awards (selected from total 44 presentations) during the closing ceremony. Prof. Geoffrey L. Smith (President of the IUMS) and Prof. Chirika Shova Tamrakar (Dean, Institute of Science and Technology, Tribhuvan University, Kathmandu) expressed their positive impression about the content, quality, and performance of the conference in their closing remarks.

We are thankful to the IUMS delegates including President Prof. Geoffrey L. Smith (Department of Pathology, University of Cambridge, England); President Elect Prof. Yuan Kun Lee (Yong Loo Lin School of Medicine, National University of Singapore, Singapore); Vice Presidents Prof. Stephen A. Lerner (Wayne State University School of Medicine, Michigan, USA) and Prof. Pierre J. Talbot (University of Quebec, Canada and organizing President of IUMS 2014 congress), and IUMS Ambassador to Asia Prof. Fusao Tomita (Hokkaido University, Japan) for their cordial support.

We would like to extend our special thanks to Prof. Suresh Raj Sharma (then Vice Chancellor of Kathmandu University) for the kind patronage of the conference. Particular thanks go to Mr. Dhurba Acharya, Mr. Radium Adhikari, Mr. Santosh Adhikari, Mr. Netra Lal Bhandari (Conference Secretary), Ms. Sulakshana Bista, Mr. Kedar Nath Dhakal, Dr. Surendra Kumar Gautam, Ms. Jyoti Giri, Mr. Khagendra Prakash K. C., Mr. Eak Raj Kadariya, Mr. Suresh Raj Kandel, Mr. Santosh Khanal, Mr. Bhuvan Khatri, Mr. Prayag Raj Kuinkel, Mr. Surendra Kumar Madhup, Mr. Rajesh Pandit (Treasurer), Mr. Tarini Prasad Paneru, Mr. Chakravarty Paudel, Ms. Sharmila Pradhan, Mr. Krishna Kumar Raut, Mr. Din Dayal Regmi, and many other students and researchers affiliated to Kathmandu University, NPI as well as NAMM, and Nanochemistry and Polymer Research Group at Tribhuvan University for their untiring efforts to make ICIDN-2012 a grand success. Thanks are due to the members of National and International Advisory Boards for their valuable feedback and support.

The financial support to the congress was generously granted by the Nepal Polymer Institute (NPI), Kathmandu University, University Grants Commission (UGC), National School of Sciences (NSS), Lainchour; Kantipur College of Medical Sciences (KCMS), Sitapaila, and National Health Care Ltd., Kathmandu.

This conference has definitely sensitized the young scientists and students from Nepal toward strengthening Research and Education in various avenues of nanomedicine, infectious diseases, virology, and antibiotic resistance. The congress is hoped to be a milestone in promoting education and researches in the field of microbiology, drug delivery, and biomedical sciences in particular fostering our endeavors in sustainable applied Nanoscience and Nanotechnology in Nepal.

We take this opportunity to thank Mr. Aninda Bose and Dr. Tobias Wasserman from Springer publication for their interest and cordial support in bringing out this special issue.

Rameshwar Adhikari
Santosh Thapa

Contents

Contributors

Rameshwar Adhikari Central Department of Chemistry, Tribhuvan University, Kirtipur, Kathmandu, Nepal; Nepal Polymer Institute, P.O. Box 24411, Kathmandu, Nepal

Lucia Baldi Ecole Polytechnique Federale de Lausanne, CH-1015 Lausanne, Switzerland

Anastasia Baranova State Research Center for Applied Microbiology and Biotechnology, Obolensk, Moscow Region, Russia, 142279

Evgenia Baranova State Research Center for Applied Microbiology and Biotechnology, Obolensk, Moscow Region, Russia, 142279

Sergey F. Biketov State Research Center for Applied Microbiology and Biotechnology, Obolensk, Moscow Region, Russia, 142279

Charlotte Boucher Department of Microbial, Biochemical and Food Biotechnology, University of the Free State, P.O. Box 399, Bloemfontein 9300, South Africa

Robert Bragg Department of Microbial, Biochemical and Food Biotechnology, University of the Free State, P.O. Box 399, Bloemfontein 9300, South Africa

Élodie Brison Laboratory of Neuroimmunovirology, INRS-Institut Armand-Frappier, Institut national de la recherche scientifique, Université du Québec, 531, Boulevard des Prairies, Laval, QC, Canada

Elke Coetsee Department of Microbial, Biochemical and Food Biotechnology, University of the Free State, P.O. Box 399, Bloemfontein 9300, South Africa

Marc Desforges Laboratory of Neuroimmunovirology, INRS-Institut Armand-Frappier, Institut national de la recherche scientifique, Université du Québec, 531, Boulevard des Prairies, Laval, QC, Canada

Galina Feduykina State Research Center for Applied Microbiology and Biotechnology, Obolensk, Moscow Region, Russia, 142279

Andrew Gdowski Department of Molecular Biology and Immunology, Graduate School of Biomedical Sciences, University of North Texas Health Science Center, Fort Worth, TX 76107, USA; Texas College of Osteopathic Medicine, University of North Texas Health Science Center, Fort Worth, TX 76107, USA

Ramaz Katsarava Institute of Chemistry and Molecular Engineering, Agricultural University of Georgia, 13 km. David Agmashenebeli Alley, 0131 Tbilisi, Georgia

Johan L. F. Kock Department of Microbial, Biochemical and Food Biotechnology, University of the Free State, P.O. Box 399, Bloemfontein 9300, South Africa

Liubov Kolombet State Research Center for Applied Microbiology and Biotechnology, Obolensk, Moscow Region, Russia, 142279

Nino Kupatadze Institute of Chemistry and Molecular Engineering, Agricultural University of Georgia, 13 km. David Agmashenebeli Alley, 0131 Tbilisi, Georgia

Alain Le Coupanec Laboratory of Neuroimmunovirology, INRS-Institut Armand-Frappier, Institut national de la recherche scientifique, Université du Québec, 531, Boulevard des Prairies, Laval, QC, Canada

Ji-Yun Lee Department of Microbial, Biochemical and Food Biotechnology, University of the Free State, P.O. Box 399, Bloemfontein 9300, South Africa

Mathieu Meessen-Pinard Laboratory of Neuroimmunovirology, INRS-Institut Armand-Frappier, Institut national de la recherche scientifique, Université du Québec, 531, Boulevard des Prairies, Laval, QC, Canada

Tamar Memenishvili Institute of Chemistry and Molecular Engineering, Agricultural University of Georgia, 13 km. David Agmashenebeli Alley, 0131 Tbilisi, Georgia

Muhammad G. Morshed Faculty of Medicine, Department of Pathology and Laboratory Medicine, BC Public Health Microbiology and Reference Laboratory, University of British Columbia, Provincial Health Services Authority, 655 West 12th Avenue, Vancouver, BC V5Z 4R4, Canada

Anindita Mukerjee Department of Molecular Biology and Immunology, Graduate School of Biomedical Sciences, University of North Texas Health Science Center, Fort Worth, TX 76107, USA; Institute for Cancer Research, University of North Texas Health Science Center, Fort Worth, TX 76107, USA

Evgeny Panfertsev State Research Center for Applied Microbiology and Biotechnology, Obolensk, Moscow Region, Russia, 142279

Carolina H. Pohl Department of Microbial, Biochemical and Food Biotechnology, University of the Free State, P.O. Box 399, Bloemfontein 9300, South Africa

Amalendu P. Ranjan Department of Molecular Biology and Immunology, Graduate School of Biomedical Sciences, University of North Texas Health Science Center, Fort Worth, TX 76107, USA; Institute for Cancer Research, University of North Texas Health Science Center, Fort Worth, TX 76107, USA

Xiao Shen Ecole Polytechnique Federale de Lausanne, CH-1015 Lausanne, Switzerland

Pavel Solov'ev State Research Center for Applied Microbiology and Biotechnology, Obolensk, Moscow Region, Russia, 142279

Chantel W. Swart Department of Microbial, Biochemical and Food Biotechnology, University of the Free State, P.O. Box 399, Bloemfontein 9300, South Africa

Pierre J. Talbot Laboratory of Neuroimmunovirology, INRS-Institut Armand-Frappier, Institut national de la recherche scientifique, Université du Québec, 531, Boulevard des Prairies, Laval, QC, Canada

David Tugushi Institute of Chemistry and Molecular Engineering, Agricultural University of Georgia, 13 km. David Agmashenebeli Alley, 0131 Tbilisi, Georgia

Wouter van der Westhuizen Department of Microbial, Biochemical and Food Biotechnology, University of the Free State, P.O. Box 399, Bloemfontein 9300, South Africa

Jamboor K. Vishwanatha Department of Molecular Biology and Immunology, Graduate School of Biomedical Sciences, University of North Texas Health Science Center, Fort Worth, TX 76107, USA; Institute for Cancer Research, University of North Texas Health Science Center, Fort Worth, TX 76107, USA

Christine Wandrey Ecole Polytechnique Federale de Lausanne, CH-1015 Lausanne, Switzerland

Nino Zavradashvili Institute of Chemistry and Molecular Engineering, Agricultural University of Georgia, 13 km. David Agmashenebelii Alley, 0131 Tbilisi, Georgia

Auger-Architectomics: Introducing a New Nanotechnology to Infectious Disease

Chantel W. Swart, Carolina H. Pohl and Johan L. F. Kock

Abstract In 2010, we developed a new imaging nanotechnology called Auger-architectomics, to study drug biosensors in nano-detail. We succeeded in applying Auger atom electron physics coupled to scanning electron microscopy (SEM) and Argon-etching to cell structure exploration, thereby exposing a new dimension in structure and element composition architecture. Auger-architectomics was used to expose the fate and effect of drugs on cells. This technology should now be expanded to diseased cells. This paper will outline the development, proof of concept, and application of this imaging nanotechnology. A virtual tour is available at: http://vimeo.com/user6296337.

Keywords Auger-architectomics · Infectious disease · Nanotechnology · NanoSAM

Introduction

Since 1982, research by our group showed that the sexual reproductive stages in yeast (asci) and sporangia of other fungi and fungus-like organisms are characterized by increased levels of mitochondrial activity when compared to vegetative reproductive structures. This is attributed to an increased energy demand needed for the development of such complex spore dispersal structures [1].

It was found that asci and other fruiting structures are selectively inhibited by various anti-mitochondrial drugs, for example, antimycin A and rotenone, as well as when oxygen was limited. This is in contrast to the vegetative cells (i.e., yeast cells and hyphae) that were less affected. This phenomenon was found to be highly

C. W. Swart · C. H. Pohl · J. L. F. Kock (✉)
Department of Microbial, Biochemical and Food Biotechnology, University of the Free State, Bloemfontein, Free State, South Africa
e-mail: kockjl@ufs.ac.za

R. Adhikari and S. Thapa (eds.), *Infectious Diseases and Nanomedicine I*,
Advances in Experimental Medicine and Biology 807,
DOI: 10.1007/978-81-322-1777-0_1,

conserved among fungi and prompted the following hypothesis: (i) That the sexual and fruiting phases of fungi are more sensitive to anti-mitochondrial drugs and other mitochondrial inhibitory environments compared to the asexual growth phases; and (ii) That sexual and other fruiting structures of fungi may be used as biosensors to screen for mitochondrial inhibitors [1]. This hypothesis has been visualized with the development of the Anti-mitochondrial Antifungal Assay (3A system), where especially pigmented yeast sexual cells are used as color biosensors to screen for anti-mitochondrial drugs. Here, changes in the color of yeast growth indicate positive hits [2].

Consequently, various 3A systems with different indicator yeasts were developed and applied by our group. Strikingly, the anti-mitochondrial activity of many well-known drugs [such as anti-cancer, antifungal, non-steroidal anti-inflammatory (NSAIDs), and anti-malarial drugs] were detected using these 3A systems. Interestingly, some of these drugs are known to pose mitochondrial liabilities in humans, while others have beneficial dual actions [3]. For example, NSAIDs may be used as effective antifungal [4, 5] as well as anti-cancer agents while anti-cancer drugs may also inhibit fungal infections [3].

It soon became apparent that biosensors may also provide valuable information regarding the holistic influence of these drugs on cells. These may be used to forecast activity, especially during further downstream tests of new drugs. Consequently, it was decided to develop a new imaging nanotechnology called Auger-architectomics, to study the effects of drugs on biosensors in nano-detail.

What is Auger-Architectomics?

This nanotechnology concerns the visualization of the atomic and 3D architecture of cells using Auger electron optics (based on Auger atom electron physics) linked to scanning electron microscopy (SEM), while etching with Argon [3, 6]. Auger-architectomics is a new and unique nanotechnology to biology and medicine and has been adapted from the field of Physics where it is used to analyze semi-conductors and the like. The challenge was to adapt this technology to fit biology and medicine [3, 6, 7]. The main adaptation hinges on the sample preparation procedure. This should assure atom and 3D structure stability while Argon nano-etching occurs during nano scanning Auger microscopy (NanoSAM; Fig. 1) and SEM visualization with an electron beam at 25 kV instead of 5 kV as per normal SEM. Sample fixation and dehydration methods that do not create sample distortions, had to be developed and optimized for NanoSAM. Dehydration regimes based on alcohol extraction procedures were used and optimized while fixation using various fixatives was included. Electron conductivity of samples through Argon-etching was assured by optimized gold sputtering.

With Auger-architectomics the surface of the gold-plated biological sample is (i) scanned in SEM mode and the surface visually enlarged. (ii) Auger atom electron

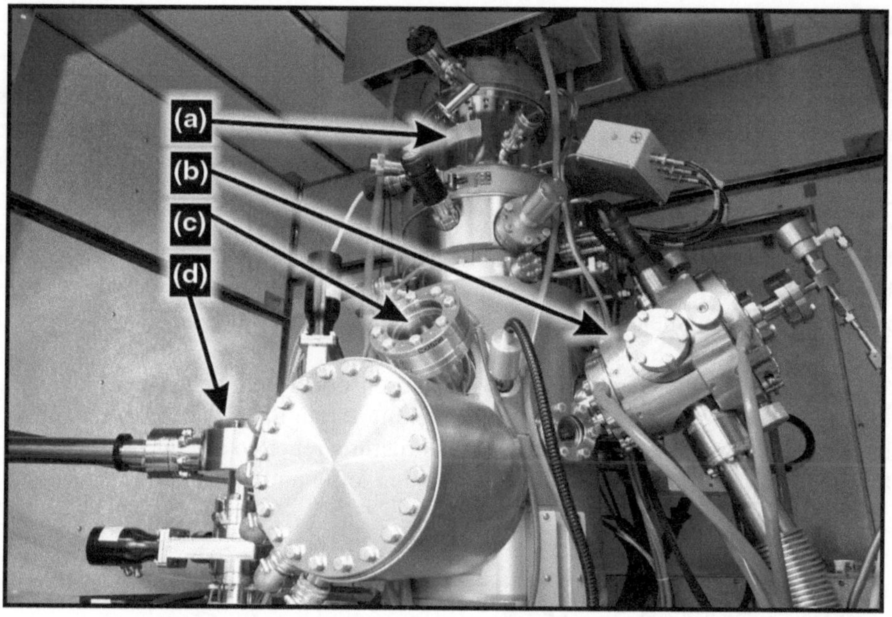

Fig. 1 A nano scanning Auger microscope photo showing (**a**) the electron gun at the top as well as the different detectors similar to that found in scanning electron microscopy (SEM), the ion gun (**b**) that uses Argon to etch the samples, the viewport (**c**) where samples can be viewed in the working chamber and the introductory chamber (**d**), where the samples are placed before entering the working chamber. Taken with permission from Kock et al. [5]

physics are applied and selected target areas on the sample surface are beamed with electrons. The incident beam ejects an electron in the inner orbital of the atom, leaving a vacancy. This is filled by an electron from an outer orbital by relaxation. Energy is released causing the ejection of an electron from the outer orbital. This electron is called an Auger electron. The amount of energy that is released is measured by Auger electron spectroscopy (AES). Since each element has a specific Auger profile, the atoms in the sample as well as their intensity/concentration can be determined. Similarly, the surface area can be screened by an electron beam eventually yielding Auger electrons that are mapped, showing the distribution of different atoms in different colors covering a surface area of predetermined size. (iii) The surface of the sample is etched with Argon at 27 nm/min, exposing a new surface of the sample that is then again analyzed as described above. In this way the 3D architecture is visualized in SEM as well as element composition mode. This nanotechnology unlocks a new field in biology where the 3D ultrastructure and element composition of the whole cell is determined while nano-"dissected" with Argon.

Proof of Concept

This was achieved by successfully linking Auger-architectomics with biosensors of *Nadsonia fulvescens* treated with high and low concentrations of fluconazole [3, 6]. Biosensor results suggest that fluconazole at high concentrations also has an anti-mitochondrial action by selectively inhibiting mitochondrial-dependant biosensor development as is indicated by the white zone in Fig. 2a. This zone contained immature spores and is characterized by a wall surrounding a transparent bubble-like structure (Fig. 2c). This was not the case in the brown zone where mature, brown-pigmented biosensors formed, each housing a fully developed spore (Fig. 2b).

Interesting results were obtained when biosensors of both zones were subjected to Auger-architectomics (Fig. 3). When etching through the fluconazole-inhibited biosensors (Fig. 3c, d), the bubble-like structure reported in Fig. 2c was confirmed. As etching started, a wall consisting of mainly carbon was recorded,

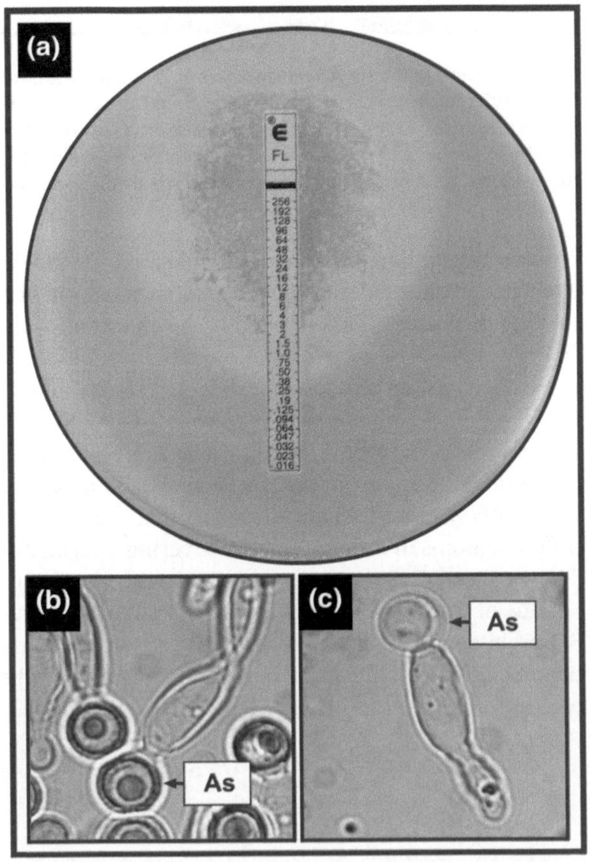

Fig. 2 The effect of fluconazole on growth and asci (As) development of *Nadsonia fulvescens*. **a** Antifungal bio-assay based on the dilution plate method showing inhibition of cell growth (*blue zone*) at high concentrations of fluconazole, inhibition of mainly ascus development (*white zone*) at lower concentrations and eventually no inhibition of sexual stage development (*brown zone*) at even lower concentrations. **b** Light micrograph of mature ascus with ascospore surrounded by spiky protuberances (isolated from *brown zone* in (**a**)). **c** Light micrograph of a bubble-like structure with less dense area positioned inside an ascus (isolated from *white zone* in (**a**)). Parts taken with permission from Swart et al. [6]

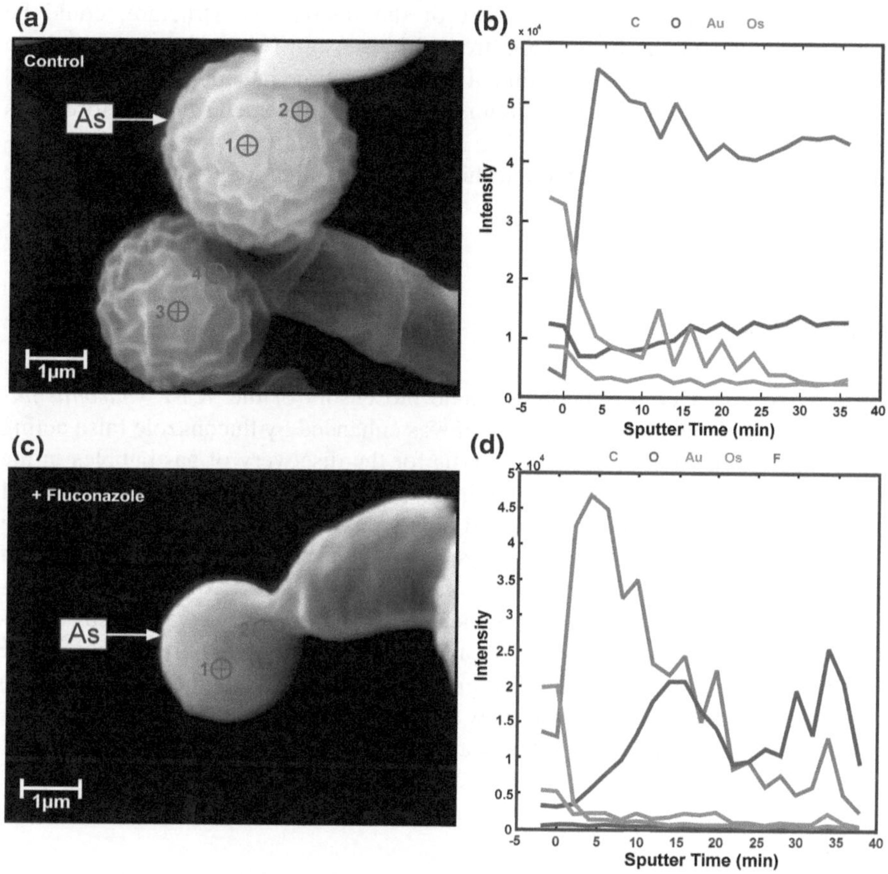

Fig. 3 Element analysis through biosensors (asci = As) during sequential etching of the yeast *Nadsonia fulvescens*. **a** Scanning electron microscopy (SEM) micrograph of two biosensors from the *brown zone*, each attached to a parental cell (Control = low fluconazole concentrations). The crossed circles show targets for element analysis. **b** A graph depicting element intensity (=concentration) over sputter time of the Control sample in (**a**) (Target 3). **c** An SEM micrograph of a smooth-walled fluconazole treated (at high concentrations) biosensor attached to a parental cell as found in the *white zone*. *Crossed circles* indicate the targets for element analysis. **d** A graph showing element intensity (=concentration) over sputter time of a fluconazole treated ascus taken from the *white zone* (Target 2). Taken with permission from Swart et al. [6]

whereafter a drastic decrease in carbon concentration occurred (Fig. 3d) when compared to the control (Fig. 3a, b). Strikingly, after etching to a depth of 945 nm (i.e., after 35 min of mining at 27 nm/min) into the biosensor, a C:O concentration ratio of 1:2 was recorded. This ratio fits the molecule CO_2. This was not surprising since inhibition of mitochondrial function of this fermenting yeast will switch on fermentation metabolism thereby increasing CO_2 production [8]. This served as the key for the unexpected discovery of gas bubbles inside brewer's and baker's yeasts.

In addition, fluorine, which is part of the fluconazole structure, could be recorded throughout the fluconazole-treated biosensor. This makes it possible to follow the distribution of this antifungal for the first time in cells. The same should be possible for other drugs and compounds that contain elements that are foreign to cells. With the above results, proof of concept was achieved for Auger-architectomics and it was clearly shown that this nanotechnology may be used to analyze biosensors in nano-detail.

Application

The discovery of CO_2 gas bubbles inside biosensors of the yeast *Nadsonia fulvescens*, where fermentation metabolism was enhanced by fluconazole (also acting as anti-mitochondrial) [3], forms the basis for the discovery of gas bubbles in the biotechnologically important fermenting yeast *Saccharomyces*. This group of yeasts is used in the brewing, wine, and bread making industries. Strikingly, using this nanotechnology, we found that a significant part of yeast cells consisted of CO_2 gas bubbles under fermentative conditions [8]. This was shown by Auger-archtectomics in SEM as well as element analysis modes and verified by transmission electron microscopy (TEM) and light microscopy. Furthermore, these bubbles drastically deformed the life giving organelles inside these cells (Fig. 4) [9]. Auger-architectomics should now be applied to all cells to determine the conserved status of these gas bubbles and the presence of other inclusions.

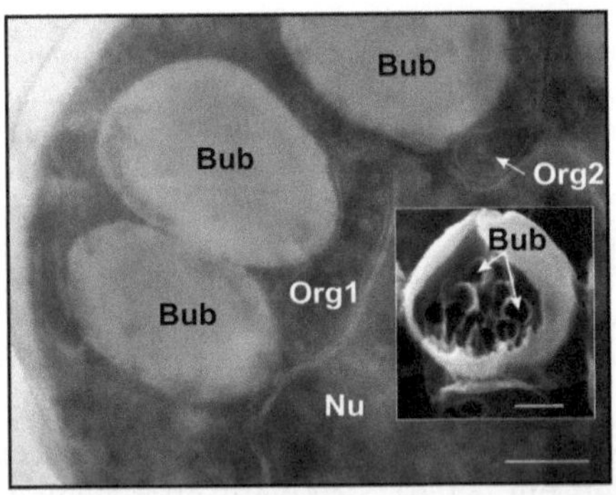

Fig. 4 Intracellular gas bubbles that deform organelles present inside fermenting brewing yeasts. Bub: gas bubbles; Org organelles; Nu: nucleus. Insert: yeast filled with gas bubbles—exposed by Auger-architectomics in scanning electron microscopy (SEM) mode. Scale bar, 200 nm. Scale bar insert, 1 μm. Taken with permission from Swart et al. [9]

Conclusions

This review highlights, in a nutshell, Auger-architectomics as a nanotechnology that is applicable to biomaterials. Not only may it be applied to disassemble biosensors and at the same time follow drug delivery and concomitant cell response on ultrastructural and element level, but may also expose new cell structures in nano-detail [10, 11]. This technology should now be applied to all cells, especially to pathogens and concomitant cell responses observed with infectious disease.

A problem with this tool concerns the sputtering and fixation steps that introduce contaminants such as gold, osmium, or phosphate to the samples prepared for the SEM part of this nanotechnology. Care should therefore be taken when interpreting results, especially regarding element composition and architecture. An ideal way of overcoming this obstacle is to use tracer elements that are foreign to the cell and not present as a mentioned contaminant. This strategy has been used successfully to track zinc inside yeast cells when grown in the presence of $ZnSO_4$ [8]. Here, zinc was used as a CO_2 gas bubble marker that "galvanized" the outer lining of gas bubbles that are present inside these cells, thereby further demarcating their respective architecture and composition. Consequently, drugs and other compounds with such "cell-foreign" probes may be easily tracked by Auger-architectomics, thereby obtaining information about their distribution and metabolic fate. This will contribute to our knowledge of, for example, cell metal detoxification processes. This approach may find application in especially nanomedicine, where gold nanoparticles attached to targeting molecules and drugs are used in target delivery.

A drawback of the 3A system is that false positive hits may be obtained by compounds that inhibit sexual reproduction processes such as meiosis and not mitochondrial activity [3]. Auger-architectomics may be applied in this case to further analyze the response of biosensor sexual cells toward these compounds to obtain a better understanding of their metabolic activity in vivo [10]. This has been demonstrated in the brewer's and baker's yeasts, where the enhancement of fermentation and therefore CO_2 production, could clearly be visualized by this nanotechnology [8]. This implies that an increase in bubble formation inside yeast cells may now also be used to detect mitochondrial inhibitors or inhibiting conditions such as anoxic environments that will inhibit respiration and favor fermentation and thus CO_2 production in fermenting fungi [8]. The in vivo response of cells toward other drugs as well as drug metabolism should now be assessed with this nanotechnology. Gas formation in general by different types of prokaryotic and eukaryotic cells should also be further researched with this technology and their exact composition and effects on cell metabolism, especially via organelle deformation, assessed. The possibility to use gas metabolism as a target to combat various diseases such as fungal infections by fermenting *Candida* should be studied using Auger-architectomics. Auger-architectomics has also successfully been applied to bacteria (see Chap. 7 of this volume and Chapter 1 of Vol. 808 of Advances in Experimental Medicine and Biology).

Acknowledgments The authors would like to thank The National Research Foundation (NRF) and The University of the Free State, South Africa, for financial support.

References

1. Kock JLF, Sebolai OM, Pohl CH, Van Wyk PWJ, Lodolo EJ (2007) Oxylipin studies expose aspirin as antifungal. FEMS Yeast Res 7:1207–1217
2. Kock JLF, Swart CW, Ncango DM, Kock Jr JLF, Munnik IA, Maartens MMJ, Pohl CH, Van Wyk PWJ (2009) Development of a yeast bio-assay to screen anti-mitochondrial drugs. Curr Drug Dev Tech 6(3):186–191
3. Kock JLF, Swart CW, Pohl CH (2011) The anti-mitochondrial antifungal assay for the discovery and development of new drugs. Expert Opin Drug Discov 6(6):671–681
4. Davis HJ, Sebolai OM, Kock JLF, Lotter AP (2009) Use of non-steroidal anti-inflammatory drugs in the treatment of opportunistic infections. Patent application no. 2009/06259
5. Kock JLF, Swart CW, Ncango DM, Pohl CH (2011) Yeast "contraceptives" also novel drugs. http://obiocon.blogspot.com/2011/05/yeast-contraceptives-also-novel-drugs.html
6. Swart CW, Swart HC, Coetsee E, Pohl CH, Van Wyk PWJ, Kock JLF (2010) 3-D architecture and elemental composition of fluconazole treated yeast asci. Sci Res Essays 5(22):3411–3417
7. Kock JLF, Swart CW (2011) A new Nanotechnology for translational medicine. http://obiocon.blogspot.com/2011/03/new-nanotechnology-for-translational.html
8. Swart CW, Dithebe K, Pohl CH, Swart HC, Coetsee E, Van Wyk PWJ, Swarts JC, Lodolo EJ, Kock JLF (2012) Gas bubble formation in the cytoplasm of a fermenting yeast. FEMS Yeast Res 12:867–869
9. Swart CW, Dithebe K, Van Wyk PWJ, Pohl CH, Swart HC, Coetsee E, Lodolo E, Kock JLF (2013) Intracellular gas bubbles deform organelles in fermenting brewing yeasts. J Inst Brew 119:15–16
10. Swart C, Olivier A, Dithebe K, Pohl C, Van Wyk P, Swart H, Coetsee E, Kock L (2012) Yeast sensors for novel drugs: chloroquine and others revealed. Sensors 12(10):13058–13074
11. Olivier AP, Swart CW, Pohl CH, Van Wyk PWJ, Swart HC, Coetsee E, Schoombie S, Smit J, Kock JLF (2013) The "firing cannons" of *Dipodascopsis uninucleata* var. *uninucleata*. Can J Microbiol 59(6):413–416

Rational Design of Antigens to Improve the Serodiagnosis of Tick-Borne Borreliosis in Central Regions of Russia

Evgenia Baranova, Pavel Solov′ev, Evgeny Panfertsev, Anastasia Baranova, Galina Feduykina, Liubov Kolombet, Muhammad G. Morshed and Sergey Biketov

Abstract Tick-borne borreliosis (Lyme disease-LD) is caused by pathogenic Borrelia spirochetes that is transmitted through bite of *Ixodes* ticks to humans and animals. In the Russian Federation, borreliosis registered with an index of 6–7 per 100,000 people annually. In reality, LD morbidity in Russia is much higher because Russian strains develop less erythematous rashes compared to North American strains, thus missed by physicians in most of the early cases, and current serology tests have insufficient sensitivity as well. The aim of this work was to improve the sensitivity and specificity of serology tests for LD in Russia using rationale-designed Borrelia antigens. It was anticipated that sensitivity of LD sero-diagnosis will be higher if antigen for test-systems are derived from a strain that is circulated in a geographical region of test application. A large portion of the Russian population lives in the Central region. Thus, effort has been made to create a serological test using antigens from Moscow region, Tula and Ul'janovsk areas. In this study we included wild strains (ultrasonic-treated spirochetes *B. garinii* H19, *B. afzelii* P1, B. *afzelii* P1H13, *B. burgdorferi s.s.* 39/40, *B. burgdorferi s.s.* B31), recombinant (expressed in E.coli DbpA, Bgp, Bbk *B. garinii*, and *B. afzelii*) antigens and some of their combinations were produced and tested against LD patients and donors serum collected in hospitals of Central regions of Russia by ELISA and Western blotting. Considering sensitivity and specificity, DbpA *B. afzelii* and DbpA *B. garinii* recombinant antigens were selected among all probed

E. Baranova (✉) · P. Solov′ev · E. Panfertsev · A. Baranova · G. Feduykina · L. Kolombet · S. Biketov
State Research Center for Applied Microbiology and Biotechnology, Obolensk, Moscow region, Russia 142279
e-mail: Givanova1@yandex.ru

M. G. Morshed
British Columbia Centre for Disease Control Public Health Microbiology and Reference Laboratory, Vancouver, BC V5Z 4R4, Canada

Department of Pathology and Laboratory Medicine, University of British Columbia, Vancouver, BC, Canada

R. Adhikari and S. Thapa (eds.), *Infectious Diseases and Nanomedicine I*, Advances in Experimental Medicine and Biology 807,
DOI: 10.1007/978-81-322-1777-0_2,

antigens for regional serology test. As long as DbpA *B. afzelii* and DbpA *B. garinii* antigens interacted with LD patient's serum in a complementary mode, it is possible to combine epitopes DbpA *B. afzelii* and *B. garinii* in a single antigen for improving sensitivity. We created recombinant fusion protein DbpA *B. afzelii/B* using dbpA genes from Russian isolates of *B. afzelii* and *B. garinii* in *E. coli.* Fusion DbpA A + G protein was then used for formulation of fast immunochromatographic serodiagnosis test (LF) in a "deep-stick" format. The trials of LF-test were conducted separately at Institute of Rheumatology Russian Academy of Medical Science (using 325 sera) and at the Borreliosis Reference Center of Ministry of Health RF (using 120 reference sera). The average sensitivity and specificity of LF-test was 80.5 and 100 %, respectively.

Keywords Lyme disease · Borrelia garinii · Borrelia afzelii · DbpA · Serology · LF-test

Introduction

Lyme disease (Lyme borreliosis) is an emerging global public health problem. Over 10,000 cases are registered in the USA, 50,000 cases in Europe, and 8,000 cases in Russia annually. The causative agent of borreliosis is *Borrelia burgdorferi sensu lato* transmissible by *Ixodes* ticks [1–4]. The group of *B. burgdorferi sensu lato* divided further into 13 species (genotypes). This classification has been based on outer surface proteins (Osp A, Osp B, Osp C), flageller gene structures, other protein profiles, metabolic activities, and genetic diversity determined through molecular typing and/or sequencing [4]. Direct detection of Lyme *Borrelia* in infected vectors, host tissues, and clinical specimens from patients include microscope-based assays, antigen detection assays, in vitro cultivation, and molecular tests. Lyme borreliosis is caused by mostly three species: *Borrelia burgdorferi* sensu stricto (s.s), *B. afzelii* ,and *B.garinii* found in Europe and Asia [5] and North America [6, 7]. Lyme borreliosis may be caused by other species of *Borrelia burgdorferi* sensu *lato* as well and can be diagnosed using both clinical and laboratory information such as *B. bissettii sp.nov.*, *B. vailaisiana*, and *B. miyamotoi* may also cause infection in humans [8, 9]. *B. afzelii* and *B. garinii* are predominantly found in Central (European) regions of the Russian Federation. Antibiotics could effectively treat Lyme disease during its early stage, however, illness may persist in host for years if not treated and the person may end up with a long-term disability. Undoubtedly, early diagnostics is necessary for treatment and to prevent Borrelia-related complications. Diagnostic strategies may vary between early and late disease manifestations and are usually determined through clinical information and serological testing. Erythema migrans is pathognomonic and does not require any further laboratory investigations. By contrast, the diagnosis of neuroborreliosis requires the assessment of serum and cerebrospinal fluid. Lyme arthritis is diagnosed in the presence of newly recognized arthritis and high-titer serum IgG antibodies against *B. burgdorferi* [10]. It was shown in Russian hospital laboratories that foreign ELISA tests for serodiagnosis of Lyme disease detected that 30–70 % of cases depend on the geographic location and stages of disease. It may be due to the fact that antigenic determinants of Borrelia could be significantly different in geographical zones and "regional" strains need to be used as antigen sources for serological testing [11]. Earlier, we showed that sensitivity of LD serodiagnosis can be increased if whole cell ultrasonic antigen for ELISA test-systems was received from strains *B. afzelii* predominantly circulating in geographical regions of test application [12]. However, such test-systems demonstrated insufficient specificity (mainly due to cross-reaction with syphilis patient's sera) and more specific recombinant antigens were made using DNA from "regional" strains. DbpAB were selected as viable alternatives to original antigens. The aim of this study was to improve the sensitivity and specificity of serology tests for LD in Russia using rationally designed Borrelia antigens considering geographical region of test application. This fast, simple, relatively cheap, and field-applicable diagnosis of LD would greatly benefit monitoring LD programs. Our focus targeted to constructing LD serological tests in an immunochromatographic format which will improve laboratory testing of LD in Russia.

Materials and Methods

Human Sera

Human sera were obtained from patients with disseminated and late stages of LD (164 patients), autoimmune disease (26 patients), syphilis (42 patients), and also from healthy blood donors who live in endemic areas (105 donors) and healthy blood donors who live in non-endemic area (100 donors). The LD patients' sera were derived from recently diagnosed and previously untreated individuals who were recruited in 2012 at the Institute of Rheumatology Russian Academy of Medical Science, Moscow. All serum specimens were obtained after blood was drawn and stored at −70 °C prior to assay. Additionally, in this study the patients' (58 patients) and donors' (62 donors) reference sera stored at Borrelioses Reference Center of Ministry of Health RF, Obolensk, were used during trial of LF serological tests.

Strains Isolation and Typing

Borrelia Cultures were obtained from intestines of ticks *I. persulcatus* and *I. ricinus*, which have been collected in Ul'yanovsk and Tula regions. Briefly, *Borrelia* were cultured in modified BSK II medium and incubated at 33 °C for 12 weeks. Spirochete grown in BSK II media were confirmed as Borrelia species through molecular testing (RFLP). Spirochaetal DNA was extracted from cultures during late log phase and subsequently intergenic spacer between 5S and 23S genes of ribosomal RNA was amplified using primers RS1(5′-CTGCGATTCGCGGAGA-3′) and RS2 (5′-TCCTAGGCATTCACCATA-3′). Amplicons were treated with endonucleases MseI and DraI and digested. DNA were run through gel electrophoresis. Finally, speciation was confirmed using banding pattern.

Whole Cell Ultrasonic Antigen

Borrelia cell grown in 0.5 l BSK II media were precipitated by centrifugation (5 min at 5,000 g). Pellet was washed 3 × with PBS, suspended in 3 ml PBS, and treated with Ultrasonic Homogenizer Bandelin Sonopuls GM 3200 using horn MS72 with the following parameters: amplitude 308 μmss, frequency 20 kHz, pulsation ON cycle 5 s, OFF cycle 30 s three times. Whole cell ultrasonic antigen was stored at −70 °C prior to assay.

Recombinant Antigens

DNA encoding selected *Borrelia* proteins were amplified by PCR from *B. garinii H19 and B. afzelii P1* genomic DNA using Taq-polymerase (Promega, USA). Gene-specific PCR primers were designed to amplify the gene coding sequence with restriction enzyme sites at the 5′ and 3′ ends. The PCR products were digested by restriction enzymes and directionally cloned into expression vector pET32b (Novagen, Madison, WI) with a six-His tag at the N terminus. Each sequence-verified expression construct was transformed into strain BL21(DE3) to produce a recombinant protein. The next synthetic oligonucleotides were used for DNA cloning and sequencing were as follows: forward AFZ 5′-aacatatgtgtagtttaacaggaaaagctag-3′ forward GAR 5′-aaaagcttggatgtggcttaacaggagaa act-3′; reverse AFZ 5′-ggaagcttttttgatttttagtttgttt tt-3′; reverse GAR 5′-aaaagctttgtagtagcagcagtgttggc-3′; T7 forward primer 5′-taatacgactcactataggg-′3; T7 reverse primer 5′-tatgctagttattgctcagcgg-3 (Novagen). PCR reaction cycles were: 95 °C\30 s; 45 °C\30 s; 72 °C\35 s; 30 cycles. The PCR reaction comprised of 1.5 mM $MgCl_2$, 0.2 mM dNTP, and 10 pM of each primer, 20 ng of total DNA of *B. garinii* H19 or *B. afzelii* P1 were used as a matrix in PCR. The PCR products were analyzed by electrophoresis in 1 % agarose gel in the presence of ethidium bromide. The target DNA fragments were cloned into vector plasmid pET32b on restriction sites NdeI-HindIII. DNA sequence was determined by the CEQtm2000XL DNA Analysis System from Beckman Coulter. Hybrid proteins were found to be synthesized in *E. coli*. For the isolation of fusion protein DbpA(A + G)-his *Borrelia,* standard methods of refolding were used. Recombinant proteins were purified by metal-chelating Sepharose CL4B. Recombinant proteins were purified with Ni-nitrilotriacetic acid resin (Qiagen, Gaithersburg, MD) and quantified with the bicinchoninic acid protein assay (Pierce, Rockford, IL), and quality was assessed by sodium dodecyl sulfate-polyacrylamide gel electrophoresis and immunoblotting according Laemmly and Towbin [13, 14]. Soluble recombinant proteins were determined by anti-his antibodies (A7058, Sigma). Purified DbpA proteins were subjected to SDS-PAGE under reducing conditions and stained with Coomassie Brilliant Blue (10 % gel) and transferred onto a nitrocellulose membrane using 12 % gel. After blocking additional protein-binding sites, proteins on the membrane were probed with positive, negative sera, and sera from patients with syphilis and goat anti-human HRP conjugate as the secondary antibody (A8794, Sigma).

ELISA

Ninety-six-well ELISA plates (Corning Inc., Corning, NY) were coated with 100 μl of *Borrelia* antigen (1–10 μg/ml) per well in coating buffer (0.1 M carbonate buffer, pH 9.2) and incubated at 4 °C overnight. After two washes (3 min each) with 200 μl of PBST (10 mM sodium phosphate, 150 mM NaCI, and 0.1 % Tween 20, pH 7.4) per well, 200 μl of blocking buffer (PBST supplemented with 5 % nonfat dry milk, BioRad) was applied to each well. The plate was shaken at 150 rpm for 1 h at 37 °C. After three washes with PBST as described above, 100 μl of human serum diluted 1:200 with blocking buffer was added to each well. The plate was incubated with shaking at 150 rpm for 1 h at 37 °C and then washed three times with PBST as before. Each well then received 100 μl of goat anti-human IgG, IgM (0.5 μg/ml) (Sigma, A8794) conjugated to horseradish peroxidase and dissolved in blocking buffer. Plates were incubated at 37 °C for 1 h with shaking at 150 rpm. After four washes with PBST for 3, 4, 5, and 6 min, respectively, 100 μl of a solution composed of the chromogen 3,3′,5,5′-tetramethylbenzidine at 0.2 mg/ml and 0.01 % hydrogen peroxide in the buffer supplied by the manufacturer (ChemBioTest, Russia) was added, and allowed 10 min to develop color. The enzyme reaction was stopped by addition of 50 μl of 1 M H_2SO_4. The optical density (OD) was measured at 450 nm with an ELISA plate spectrophotometer model Pikon Uniplan (Russia). The cutoff OD value was defined as the mean OD plus 3 standard deviations (SDs) for 7 serum samples collected from patients of a hospital in Astrakhan' (where Lyme disease is not endemic). All the samples were assessed blindly in duplicate and repeated twice. Mean OD values from duplicate results were reported. OD values of individual samples never varied more than 5 % [3].

Formulation of LF-Test

To formulation of LF-test antigen was adsorbed on paper and protein G (or anti-human IGG) was conjugated with colloidal gold nanoparticles. For preservation of antibodies binding activity "fusion DbpA" was adsorbed on nitrocellulose paper in presence of protease inhibitors and drying-protectors. Developer reagents—conjugates protein G with gold nanoparticles with size 5, 20, or 40 nm (according Atomic Force Microscopy data) were preliminary tested concerning antibodies binding activity and visualization accuracy. As results, a "deep-stick" tests composing from nitrocellulose strip containing sample pad with dried 40 nm colloidal gold-labeled protein G and test-line with dried capture antigen ("fusion DbpA") were fabricated.

Results

Determination of Sensitivity and Specificity of Cell Antigens Based on Regional Borrelia Strains by ELISA

The strains from ticks collected in Central region of Russia (Moscow, Tula and Ul'janovsk areas) were used for whole cell antigen preparation. *Borrelia* whole cell ultrasonic antigens were received from *B. garinii* H19, *B. afzelii* P1, *B. afzelii* H13, *B. burgdorferi s.s.* 39/40, and *B. burgdorferi s.s.* B31 strains and tested with a patient's and donor's human sera.

It was shown that depending on the strains used, sensitivity for detecting by ELISA in disseminated and late stages of LD varied from 64 % (*B. burgdorferi* 39/40) to 88.0 % (*B. afzelii* P1). All strains demonstrated acceptable level of specificity with donors' sera and high-level cross-reaction with sera from syphilis patients (Table 1). Besides, the different immune reactivity with LD patients' pool sera in immunoblot for the foreign and Russian *Borrelia* strains was observed (Fig. 1). The more intensive immunoreactions with LD patients' pooled sera for Russian strains *B. garinii* H19 and *B. afzelii* P1 in the areas 40–70 and 34–18 kDa were detected.

This part of the study demonstrated that ELISA test systems for serodiagnosis of LD constructed using antigen sources corresponded to the geographical region of application demonstrated better sensitivity. These results indicated a possibility to increase sensitivity of LD serodiagnosis by using *Borrelia* strains circulated in Central region of Russia as an antigen source.

Determination of Sensitivity and Specificity of Recombinant Antigens Received Using DNA from "Regional" Strains

Since whole cell antigens demonstrated insufficient specificity (mainly via cross-reaction with syphilis pathients sera) the more specific recombinant antigens were selected as a viable alternatives to the original whole cell antigens.

We cloned and expressed genes *bbk*, *bgp*, and *dbp*a from regional strains of *B. afzelii* P1 and *B. garinii* H19 in *E. coli*. Recombinant proteins were tested through ELISA using LD patient's sera (dissemination and late stages) from European region of Russia (Table 2).

Additionally the *Borrelia* recombinant antigens were tested with LD patients' sera collected in different areas of Central region—Moscow, Tula, and Ul'janovsk.

Table 1 Interaction of various *Borrelia* whole cell antigens with patients' and donors' serum collected from Central Region of Russia

Patients sera	Borrelia antigens				
	B. afzelii H13 (%)	*B. afzelii* P1 (%)	*B. garinii* H19 (%)	*B. burgdorferi* 39/40 (%)	*B. burgdorferi* B31 (%)
LD patients ($n = 50$)	82.0	88.0	68.0	64.0	68.0
Healthy blood donors ($n = 100$)	10.0	8.0	8.0	6.0	8.0
Autoimmune disease patients ($n = 20$)	15.0	15.0	10.0	10.0	15.0
Syphilis patients ($n = 40$)	65.0	65.0	62.5	60.5	65.0

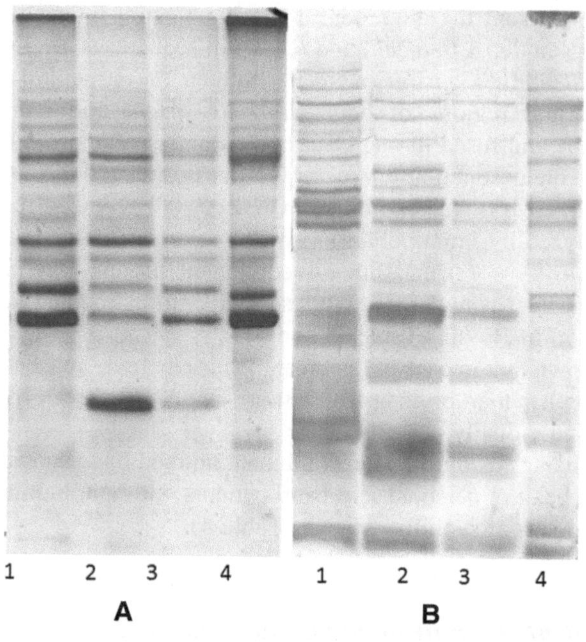

Fig. 1 SDS-PAGE and Western blot of whole cell ultrasonic antigens *B. garinii* H19, *B. afzelii* P1, *B. burgdorferi* 39/40, and B.burgdorferi B31 against pooled positive LD patient's sera. *A* SDS PAAG electrophoresis, *B* immunobloting against pooled positive LD patient's sera. *1 B. afzelii* P1, *2 B. garinii* H19, *3 B.burgdorferi* 39/40, *4 B.burgdorferi* B31

The more frequent immunoreactions of LD patients' sera from Moscow and Tula areas for DbpA *B. afzelii* were observed (66 %). Meanwhile only for 17 % of these sera reacted with DbpA *B. garinii.* The reactivity of sera from Lyme Borreliosis patients of Ul'janovsk regions with DbpA *B .afzelii* consist of 53 % and for DbpA *B. garinii*—23 %. It is important that DbpA *B. afzelii* and DbpA *B. garinii* antigens interacted with LD patient's sera in a complementary mode that creates possibility to improve sensitivity by combining DbpA *B. afzelii* and *B. garinii* in

Table 2 The sensitivity of the ELISA serology on a base of recombinant antigens for LD patient's sera (dissemination and late stages) from European Region of Russia

Lyme disease patients' sera (n = 164)	*Borrelia* antigens				
	B. afzelii Strain H13 (%)	*B. afzelii* DbpA (%)	*B. garinii* DbpA (%)	*B. afzelii* Bbk (%)	*B. afzelii* Bgp (%)
Positive	71.9	73.8	24.4	11.6	15.2
Equivocal	15.2	16.4	12.8	3.0	6.0
Negative	12.9	9.8	62.8	85.4	79.8

single antigen. The analyses of dbpA *B. afzelii* and *B. garinii* sequences (Fig. 2) indicate that Russians' strains have essential variability of dbpA sequences similar or more to described European isolates.

As we can see from the alignment, there is essential interspecies/strains heterogeneity between dbpA *B. afzelii* and *B. garinii* sequences. To combine epitopes DbpA *B. afzelii* and *B. garinii* in single antigen the structural parts of genes encoding important regions of dbpA *Borrelia afzelii* and *garinii* have been obtained by PCR and fused in one ORF under control of a strong promoter. As a result, fusion protein containing structural parts of genes dbpA *B. afzelii* and *B. garinii*—DbpA(A + G) was expressed in *E. coli*. All received recombinant proteins were analyzed by immunoblotting with pooled LD patients' sera (Fig. 3).

Fusion DbpA (A + G) and other recombinant proteins were comparatively tested though ELISA using LD patient's sera (dissemination and late stages) from European Region of Russia (Tables 3 and 4).

So it was shown that in case of fusion antigen, sensitivity reaches to 84.1 % that is essentially higher than observed in case of DbpA *B. afzelii* (73.8) or *DbpA B. garinii* (24.4) individually.

ELISA test using fusion DbpA recombinant antigen had a specificity of 100 % with sera from healthy persons, sera from patients with autoimmune disease, and 96.7 % with sera from syphilis patients (Table 4).

Development of a Immunochromatographic Test for Serodiagnosis of LD Based on "Fusion DBPA"

Fusion DBPAG protein was used for formulation of fast immunochromatographic serodiagnosis test in "deep-stick" format. On a base fusion antigen DbpAG experimental test was developed (Fig. 4).

The trials of deep stick were conducted with LD patient's sera from various disease stages. Deep stick was interacted with reference sera (dissemination and late stages) with sensitivity in a range 80.5 %. The test had a specificity of 100 % (no cross-reactivity was found with sera from healthy persons and syphilis patients). During clinic trials the possibility of novel recombinant protein to identifying specific anti-*Borrelia* IgG was demonstrated.

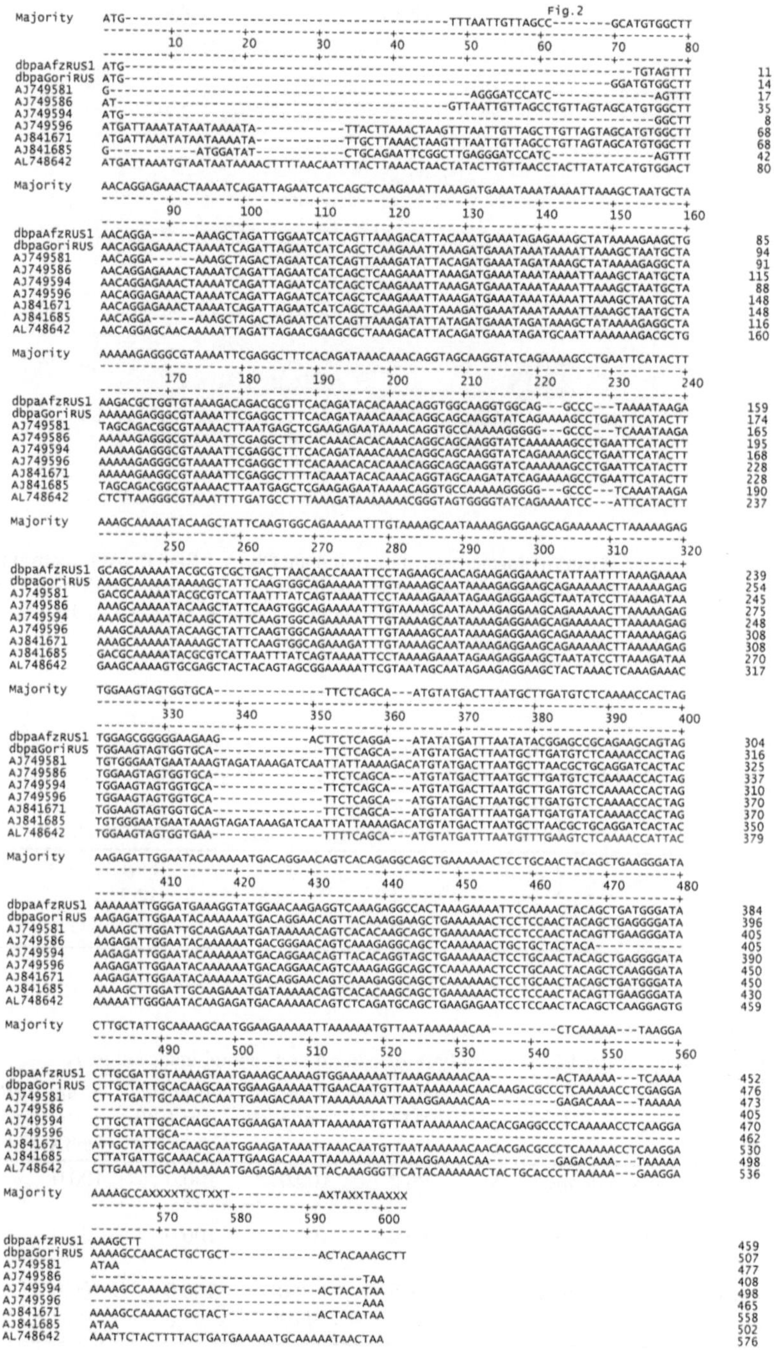

Fig. 2 Alignment of dbpa nucleotide sequences from *B. afzelii* and *B. garinii* strains isolated in different European geographic regions, including Russia

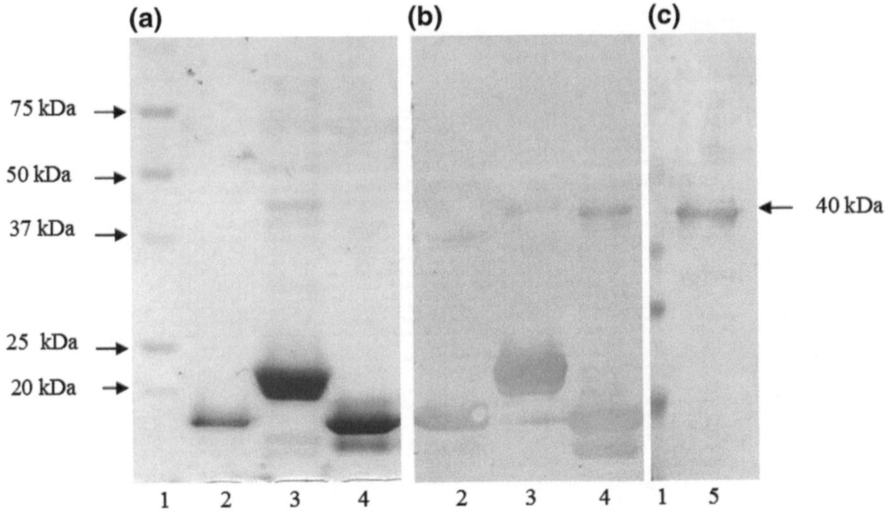

Fig. 3 SDS-PAGE and Western blot analysis of recombinant DbpA. **a** SDS PAAG electrophoresis: 1- MW markers; 2 – DbpA B.afzelii P1; 3 – DbpA B.garinii H19; 4 – DbpA B.burgdorferi B31 **b** and **c** - immunobloting with pooled positive LD patient's sera: 2 – DbpA B. afzelii; 3 – DbpA B. garinii; 4 – DbpA B. burgdorferi B31; 5 - DbpA (A+G)

Table 3 The sensitivity of the ELISA serology on a base of natural and recombinant antigens with LD patients' sera (dissemination and late stages) from European Region of Russia

Lyme disease patients' sera (n = 164)	*B. afzelii* Strain H13 (%)	*B. afzelii* DbpA (%)	*B. garinii* DbpA (%)	*B. afzelii/B. garinii* fusion DbpA (%)
Positive	71.9	73.8	24.4	84.1
Equivocal	15.2	13.4	12.8	12.2
Negative	12.9	12.8	62.8	3.7

Table 4 Specificity of the ELISA on a base of natural and recombinant antigens

Patients' sera	*B. afzelii* H13 (%)	*B. afzelii* DbpA (%)	*B. garinii* DbpA (%)	*B. afzelii/B. garinii* fusion DbpA (%)
Healthy blood donors endemic areas (n = 105)	96.2	100.0	100.0	100.0
Healthy blood donors non-endemic areas (n = 100)	99.0	100.0	100.0	100.0
Autoimmune disease patients (n = 26)	88.5	100.0	100.0	100.0
Syphilis patients (n = 42)	38.1	95.2	97.6	96.7

Fig. 4 Deep stick for detection of anti-Borrelia antibodies

Conclusion

We aimed to develop a serological test using local strains from the Central region of Moscow since a large portion of the Russian population lives in this area. We isolated a number of strains from ticks collected from Moscow and Ul'janovsk area and from the skin biopsy of patients with clinical diagnosis of Lyme diseases. The strains from ticks (*B. afzelii* P1, *B. afzelii* У6 M, *B. afzelii* H13, *B. garinii* H19) and from patients (*B. garinii* Siu, *B. garinii* Kol) were isolated using standard methods and confirmed by molecular methods such as RFLP and used as antigen sources. We searched antigenic spectrums of selected isolates. All diagnostically significant bands with molecular weights 93, 80, 66, 45, 41, 39, 37, 34, 31, 28, 25, 21 kDa are well represented in our antigen preparation of those isolates. Ultrasonic cell lysates of these strains were used as antigenic basis in experimental immunochemical test-systems produced "in house." However, false-positive result levels that were relatively high was observed. Therefore, genes dbpA, bgp, and bbk were cloned and appropriate recombinant proteins were expressed using templates DNA from isolates *B. garinii* and *B. afzelii*.

As results, recombinant antigens expressed in *E. coli*—DbpA, Bgp, Bbk *B. garinii* and *B. afzelii*, and some of their combination were produced and tested by ELISA using LD patient's and donor's serum collected in hospitals of Central regions of Russia. Taking into consideration that DbpA *B. afzelii* and DbpA

B. garinii recombinant antigens demonstrate higher specificity compared to ultrasonic cell lysates, they were targetted for the specific population (regional). Decorin-binding *Borrelia* protein DbpA represents a lipoprotein, which participates in spirochete adhesion. High titer of antibodies to DbpA among infected people makes this protein suitable for serological diagnosis of Lyme disease [15, 16]. The genes encoding DbpA synthesis from *Borrelia* species are structurally highly heterogeneous. Analysis of dbpA sequences from European isolates *B. burgdorferi s.s., B. afzelii, B. garinii*, and human pathogenic genospecies A14S revealed five distinct DbpA groups. Group I comprises *B. burgdorferi* s.s. and group II *B. afzelii, B. garinii* is divided into groups III and IV, whereas A14S strains form group V [17]. We also found that Russian isolates demonstrates essential variability of dbpA sequences. Hence, it is important that DNA encoding selected *Borrelia* proteins should be amplified by PCR from genomic DNA *of B. afzelii and B. garinii* strains circulated in geographical regions of test application. As long as DbpA *B. afzelii* and DbpA *B. garinii* antigens interacted with LD patient's sera in a complementary mode, it is possible to combine epitopes DbpA *B. afzelii and B. garinii* in single antigen for improving the sensitivity and DbpA *B. afzelii/B. garinii* (DbpA A + G) recombinant fusion protein were created [18]. Fusion DBPA G protein was used for formulation of serodiagnosis test in "deep-stick" format (rapid point of care test). The trials of test were conducted separately in the Borreliosis Reference Center of Ministry of Health RF and Institute of Rheumatology Russian Academy of Medical Science. The average sensitivity and specificity of test consisted 80.5 and 100 % appropriately. This test was found to be useful as a screen test on LD and could be accurately performed and interpreted by minimally trained healthcare workers within 20 min.

Acknowledgments This work was performed within framework of the International Science and Technology Center Project #2738.

References

1. Tugwell P, Dennis DT, Weinstein A, Wells G, Shea B, Nichol G, Hayward R, Lighfoot R, Barker P, Steere AC (1997) Clinical guideline part 2: laboratory evaluation in diagnosis of Lyme disease. Ann Intern Med 127:1109–1123
2. Steere AS, Coburn J, Gilckstain L (2004) The emergence of Lyme disease. J Clin Investig 113(8):1093–1101
3. Liang FT, Steere AS, Marques AR, Johnson BJB, Miller JN, Philipp MT (1999) Sensitive and specific serodiagnosis of Lyme disease by Enzyme-Linked Immunosorbent Assay with a peptide based on an immunodominant conserved region of *Borrelia burgdorferi* Vls. Eur J Clin Microbiol 37:3990–3996
4. Wang G, Alje P, van Dam AP, Schwarz I, Dankert J (1999) Molecular typing of *Borrelia burgdorferi* sensu lato: taxonomic, epidemiological, and clinical implications. Clin Microb Rev 12(4):633–653

5. Baranton G, Postic D, Saint Girons I, Boerlin P, Piffaretti JC, Assous M, Grimont PA (1992) Delineation of Borrelia burgdorferi sensu stricto, Borrelia garinii sp. nov., and group VS461 associated with Lyme borreliosis. Int J Syst Bacteriol 42:378–383
6. Morshed MG, Scott JD, Fernando K, Geddes G, McNabb A, Mak S, Durden LA (2006) Distribution and characterization of *Borrelia burgdorferi* isolates from Ixodes scapularis and presence in mammalian hosts in Ontario, Canada. J Med Entomol 43(4):762
7. Burgdorfer W, Anderson JF, Gern L, Lane RS, Piesman J, Spielman A (1991) Relationship of *Borrelia burgdorferi* to its arthropod vectors. Scand J Infect Dis Suppl 77:35–40
8. Wang G, van Dam AP, Le Fleche A, Postic D, Peter O, Baranton G, de Boer R, Spanjaard L (1997) Dankert J Genetic and phenotypic analysis of *Borrelia valaisiana* sp. Nov. (Borrelia genomic groups VS116 and M 19). Int J Syst Bacteriol 47:926–932
9. Mun J, Eisen RJ, Eisen L, Lane RS (2006) Detection of a *Borrelia miyamotoi* sensu lato relapsing-fever group spirochete from *Ixodes pacificus* in California. J Med Entomol 43:120–3. PubMed doi: 10.1603/0022-2585(2006)043[0120:DOABMS]2.0.CO;2
10. Huppertz HI, Bartmann P, Heininger U, Fingerle V, Kinet M, Klein R, Korenke GC, Nentwich HJ (2012) Rational diagnostic strategies for Lyme borreliosis in children and adolescents: recommendations by the Committee for Infectious Diseases and Vaccinations of the German Academy for Pediatrics and Adolescent Health. Eur J Pediatr 171(11):1619–1624
11. Nikolaenko VV, Korenberg EI, Vorobjeva NN, Roggenbruck D (2001) Using micromodification indirect IFA for serodiagnosis testing ixodi-born borreliosis. Med Parasitol 4:20–26
12. Shtannikov AV, Baranova EV, Evsegneev SI, Reshetnjak TV, Repolovskaya TV, Prilipov AG, Kolosova NV, Naumov RL, Biketov SF, Ananjeva LP (2001) Isolation and investigation of properties *Borrelia burgdorferi* sensu lato strains from clinical samples and ticks. In: Materials of seminar clinical perspective in infectology. "Ticks and parasitosis infections", 1:35–38
13. Laemmly UK (1970) Cleavage of structural proteins during the assembly of the head of Bacteriophage T4. Nature 227:680–685
14. Towbin H, Staehelin T, Gordon J (1979) Electrophoretic transfer of proteins from polyacrylamide gels to nitrocellulose sheets: procedure and some applications. Proc Nat Acad Sci USA 76:4350–4354
15. Ulbrandt ND, Cassatt DR, Patel NK, Roberts WC, Bachy CM, Fazenbaker CA, Hanson MS (2001) Conformational nature of the *Borrelia burgdorferi* decorin binding protein A epitopes that elicit protective antibodies. Infect Immun 69(8):4799–4807
16. Cassatt DR, Patel NK, Ulbrandt ND, Hanson MS (1998) DbpA, but not OspA, is expressed by *Borrelia burgdorferi* during spirochetemia and is a target for protective antibodies. Infect Immun 66(11):5379–5387
17. Schulte-Spechtel U, Fingerle V, Goettner G, Rogge S, Wilske B (2006) Molecular analysis of decorin-binding protein A (DbpA) reveals five major groups among European *Borrelia burgdorferi* sensu lato strains with impact for the development of serological assays and indicates lateral gene transfer of the dbpA gene. Int J Med Microbiol 296(40):250–266
18. Panfertsev E, Mochalov VV, Baranova EV, Kolosova NV, Solov'ev PV, Shtannikov AV, Prilipov AG, Morshed MG, Biketov SF (2009) Design of a novel technology for *Borrelia* DbpA protein construction using recombinant *E. coli* strains-producers for serological diagnostics of tick-born borreliosis. In: Materials of 12th SAC seminar combating global infections, Irkutsk, p 54

Ayurvedic Bhasmas: Overview on Nanomaterialistic Aspects, Applications, and Perspectives

Rameshwar Adhikari

Abstract In this paper, we present an overview of Ayurvedic Bhasmas as nanomedicine of herbo-metallic and mineral origin with particular attention to their structural aspects. We find, the Bhasmas as nanomedicines may offer a huge potential for designing new drugs employing the concept of nanotechnology. Thus, the standardization of fabrication process of these formulations is a crucial issue to be addressed. The structure and effectiveness of the Bhasmas as drugs depend largely on their processing history. Bhasmas are generally safe drugs for human beings in spite of the presence of seemingly toxic elements and compounds as indicated by recent studies using modern analytical techniques. Nevertheless, more systematic nanomaterialistic investigations on Bhasmas are recommended for gaining the complete and reliable composition-processing-structure-effectiveness picture of these drugs.

Keywords Ayurveda · Bhasma · Nanomedicine · Heavy metal toxicity

Introduction

Charak Samhita, the ancient work compiled by Maharshi Charaka and his followers some 2000 years ago is the basis of Ayurveda (or the Science of Life). A part of Ayurveda is concerned with the therapeutic application of metal–mineral formulations called as Bhasmas (literally meaning ashes). Different kinds of Bhasmas have been used in Ayurvedic medicine, viz., Swarna Bhasma (gold),

R. Adhikari (✉)
Central Department of Chemistry, Tribhuvan University, Kirtipur, Kathmandu, Nepal
e-mail: nepalpolymer@yahoo.com

R. Adhikari
Nepal Polymer Institute, P. O. Box 24411, Kathmandu, Nepal

R. Adhikari and S. Thapa (eds.), *Infectious Diseases and Nanomedicine I*, Advances in Experimental Medicine and Biology 807,
DOI: 10.1007/978-81-322-1777-0_3, © Springer India 2014

Tamra Bhasma (copper), Rajata Bhasma (silver), Makardwaja (an alloy of gold, sulfur, and mercury), Loha Bhasma (iron), Jasada Bhasma (zinc), Mandura Bhasma (iron oxide), etc., are some popular examples [1]. The ashes of minerals are also frequently used in Ayurvedic Bhasma medicines. The example of this category includes Coury (cowry shells), Abrakha (mica), Muga (coral), Sankha (conch shell), Sipi (oyster shell), Moti (pearl shells), etc., Bhasmas. All the Bhasmas are fabricated using traditional Ayurvedic techniques called *Bhasmikarana* [1–8]. The techniques are easy to handle and do not require pre-knowledge of sophisticated equipments. Thus, the Ayurveda system is economical and has no potential health hazards [9].

The use of herbs and herbo-metal preparations is strongly advocated in Ayurvedic systems of medicine [1–13], in which the particles used have a diameter of about 10–50 nm [5–7, 10, 13]. It is well known that the tools that can safely operate inside the living cells with metal-based or carbon-based materials up to 100 nm in dimension are termed as nanomedicines. In this respect, it is interesting to note that various metals such as zinc, iron, calcium, copper, mercury, gold, and silver (and their compounds) in particulate forms have been used both in traditional Ayurvedic system and modern nanomedicines. For instance, Jasada Bhasma derived from zinc has been used for treatment of diabetes, arthritis, and tuberculosis. Likewise, diseases such as anemia, diabetes, and rheumatism have been cured by using iron-based formulations while muscles wasting, nerve disorders, and brain diseases are cured by silver-based Rajata Bhasma [9, 13–15]. Tamra Bhasma (copper ash) has been long used for curing acidity, tuberculosis, and cirrhosis (that may lead to cancer in later stage). Similarly, rheumatoid and arthritis are being handled by Swarna Bhasma (gold ash).

In the context of growing worldwide interest on alternative medicine and the increasing resistance of microbes against refined medicaments, Ayurvedic medicines have attracted the attention of therapists worldwide.

The aim of this paper is to provide an insight into the correlation between modern nanomedicine and Ayurvedic Bhasmas in terms of structural characterization, biocompatibility, and toxicology of the Bhasma drugs. We will further briefly introduce the perspectives of these nanodrugs in our society.

Bhasmas as Nanomedicine and Their Physicochemical Characterization

Nanomedicine Fabrication and Characterization Aspects

The fabrication procedures of Bhasma formulations closely correspond to the "top-down" approach of nano-objects synthesis. The number of different steps employed, nature of the additives used, time and temperature of the treatment, etc., have tremendous impacts on the morphology and efficiency of Bhasma particles [11–13, 16, 17].

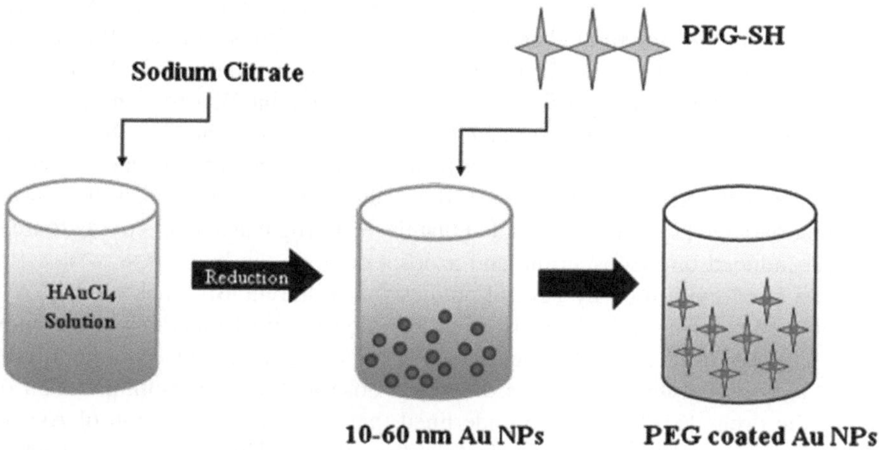

Fig. 1 Schematic representation of preparation of PEG-coated gold nanoparticles [14]

Let us begin our discussion with modern nanoparticles which are used for the preparation of nanodrugs through a wide range of chemical modifications. The nanoparticles (NPs) derived from such metals as gold (Au), silver (Ag), and copper (Cu) have been advocated for their activity in the human bodies to fight against several chronic diseases also in modern medicines [15, 18–20]. For instance, gold nanoparticles (AuNPs) have great therapeutic potential to fight against cancer; and these act as special drug delivery vehicles that can be synthesized by facile routes such as using gold containing salts [14, 15, 18]. The method is, however, entirely different from the Ayurvedic route of Swarna Bhasma preparation.

The preparation of AuNPs, 10–60 nm in diameter prepared by refluxing chloroauric acid ($HAuCl_4$ $4H_2O$) with sodium citrate (i.e., formed due to reduction of gold ions by citrate ions) is shown in Fig. 1 [14]. Addition of polyethylene glycol (PEG) solution to the resulting mixture yields the PEG-coated gold nanoparticles. Several unique physicochemical properties of gold nanoparticles make them potential candidates for targeted drug delivery vehicles in cancer therapy.

There have been many attempts to demonstrate that nanomaterials have great potential to solve several medical problems facing mankind these days. Attention has been focused on the effects of nanoparticles that enter the body accidentally. In contrast, much lesser attention has been paid toward the toxicology of nanoparticles that are used for biomedical applications, such as drug delivery or imaging, in which the nanoparticles are deliberately incorporated into the body. It was found that nanoparticles can stimulate and suppress the immune responses. Further, their compatibility with the immune system is largely determined by their surface chemistry and structures. Thus, the modification of these factors may contribute to reduce the toxicity of nanoparticles and hence make them useful platforms for drug delivery [15]. This issue is concerned with Bhasmas, which we will briefly discuss in the following paragraphs.

Recently, several Bhasmas have been investigated using modern analytical tools aiming at the structural characterization of these drugs and their functioning inside the living bodies via in vivo tests in different animals [1–7, 17–19, 21]. Bhatia et al. evaluated the protective effect of Abrakha Bhasma on spermatogenesis in the rat testes deteriorated then by action of heat. The histological analysis of *sukravaha srotomula* (testes) of albino Wistar rat was carried out to investigate the potency of the Ayurvedic drug in preventing the heat-induced testicular damage [2]. It was concluded that the test drug can act effectively to treat the heat-induced oligozoospermia and azoospermia also in humans.

It has been confirmed by several studies that different Bhasmas used in Ayurvedic medicines do not adversely affect the physiological functions of the body when properly used as medicament. In this regard, Sarkar and coworkers have briefly highlighted the importance of proper dose and way of administration of Ayurvedic drugs [3]. It has been underlined that improper application of Ayurvedic Bhasmas may lead even to the death of the patients. This fact is true, however, also for modern synthetic drugs. We will discuss the structural aspects of the Bhasma nanodrug exemplified by two different formulations (viz., Tarakeshwara Rasa and Mandura Bhasma); see the discussions below.

Tarakeshwara Rasa [17] comprises different Bhasmas (such as Abhrakha, Loha, Vanga, and Rasa Sindhura) and is generally administered to diabetes mellitus patients [17]. Recently, Tarakeshwara Rasa was prepared using standard protocols and studied by various methods. The results from the X-ray diffraction (XRD) are shown in Fig. 2a. Based on XRD and energy dispersive X-ray (EDX) analyses, the presence of different elements such as tin, iron, aluminum, mercury, calcium, manganese, and magnesium in combined forms was confirmed in this drug. SEM study revealed that the compound comprises the agglomerated particles, 0.5–2 μm in diameter, Fig. 2b.

Results on a similar line were reported recently by Jagtap et al. and Wadekar et al. on Tamra (copper) Bhasma [4, 5], Unni et al. on Annavedi Sinduram [6], Mulik and Jha on Mandura Bhasma [7], and Mohaptra and Jha on Swarna Makshika Bhasma [19]. Wadekar et al. found that the particle size of the drug was bigger than that in the standard sample, which was explained as the result of the repeated calcination procedures leading to agglomeration of the smaller copper oxide particles. In contrast, reduction in the particle size of the drug was claimed compared with that of the original material by Unni et al. [6]. One of the most important conclusions that can be drawn from the results of recent works was on the effect of drug processing history on the particles size, and hence on the final activity of the drug.

The nanostructural basis of Bhasma medicine can be illustrated further by Mandura Bhasma, an iron-based preparation used in therapeutics of anemia, jaundice, poor digestion, edema, skin diseases, etc. [7]. The chemical analysis showed that the Mandura in the raw form contained Fe_2SiO_4 while the Bhasma contained Fe_2O_3 and SiO_2 with the grains 200–300 nm in diameter organized in uniform fashion (Fig. 3b), much smaller than the micron-sized particles in the raw Mandura (Fig. 3a).

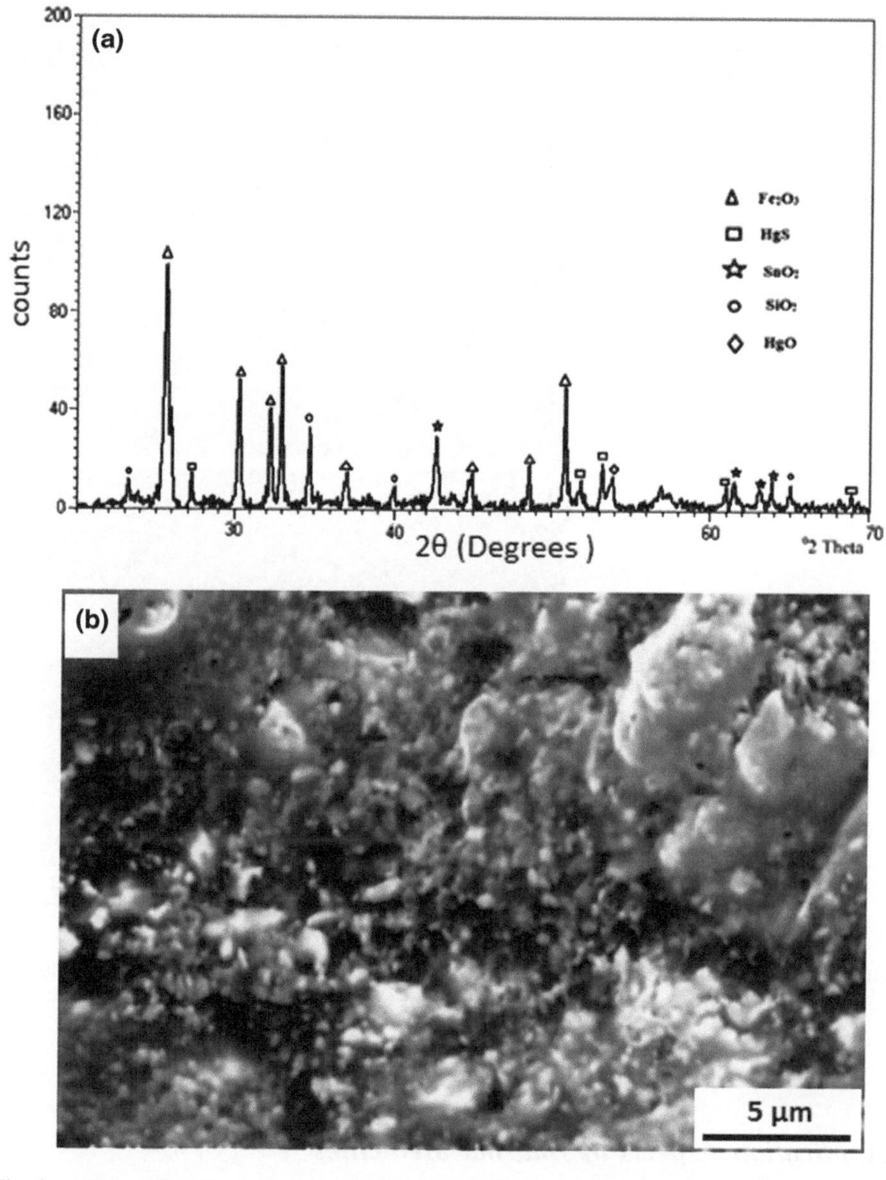

Fig. 2 **a** X-ray diffraction pattern, and **b** scanning electron microscopy image of Tarakeshwara Rasa [17]

In summary, the nanostructured morphology of Bhasmas of various types was thus demonstrated by different techniques. It should be emphasized that the morphology of various Bhasmas described in the literature, in particular with respect to texture and size of the particles, remains still contentious and requires

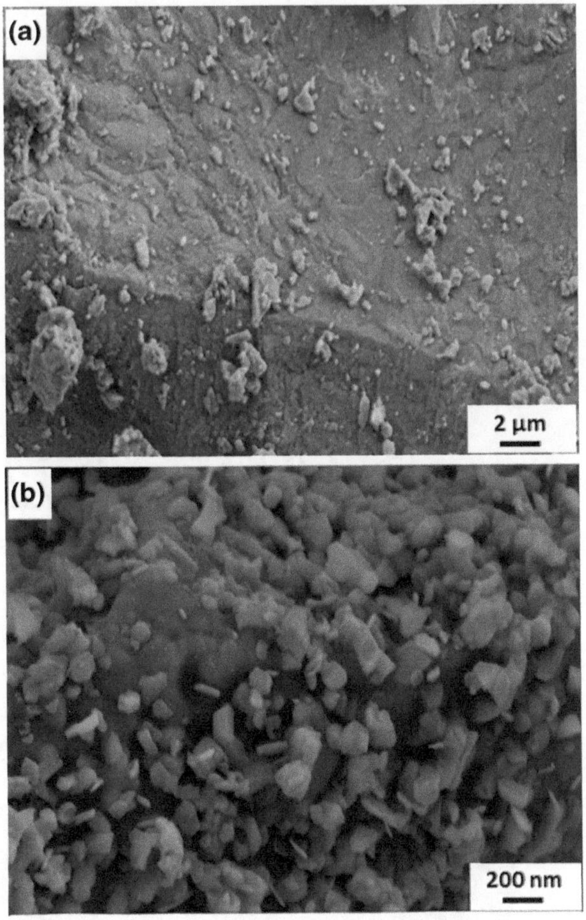

Fig. 3 Scanning electron micrographs of **a** Mandura, and **b** Mandura Bhasma [7]

more systematic investigation. Nevertheless, it is established that Bhasmas are nanodrugs and have therapeutic potential as we will describe in the next section briefly.

Application Aspects of Bhasma Medicines

So far we have reviewed the materials scientific aspect of selected Bhasmas with particular attention to their nanoscopic morphology. In the application aspects of the drugs, there are several therapeutic requirements to be met in order that the given formulation becomes a useful drug. The two crucial aspects are the compatibility with body cells and fluids, and their action on the targeted objects in the body. Here, in order to shed light on the biocompatibility and effective action of

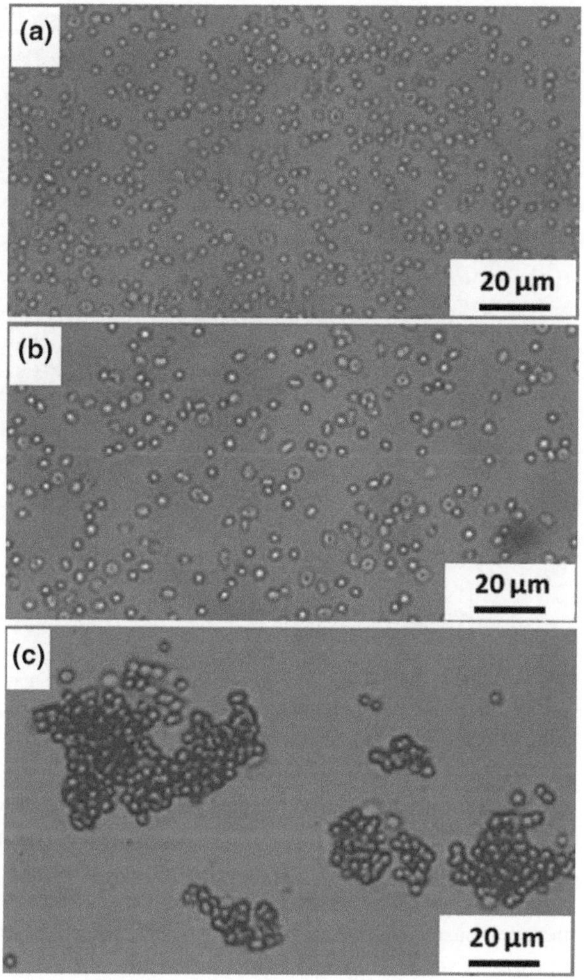

Fig. 4 Aggregation of RBC by incubation of **a** Swarna Bhasma (*Gold ash*), **b** normal saline (*negative control*), and **c** polyethylene imine, PEI (*positive control*) [11]

Ayurvedic Bhasmas, we present two examples: blood compatibility study of Swarna Bhasma [12] and cytotoxicity study of Arogyavardini Vati [13].

In vitro cytotoxicity studies of Swarna Bhasma (gold) particles, approximately 30 nm in diameter, were carried out on L929 fibroblast cells using 3-(4,5-dimethythiazol-2-yl)-2,5-diphenyl tetrazolium bromide (MTT) assay following the recommended ISO protocols [11]. For instance, the aggregations of the human blood cells subjected to interaction with the Bhasma nanoparticles, are shown in Fig. 4 for RBCs [11].

No aggregation of the blood cells was observed upon incubation of the Swarna Bhasma even at a high interaction ratio of 10 mg/ml. During the experiment, polyethyelene imine (PEI) used as positive control showed aggregation, whereas the normal saline used as negative control did not (see Fig. 4). Similar results were obtained for tests with WBCs and platelets as well as with the hemolytic property

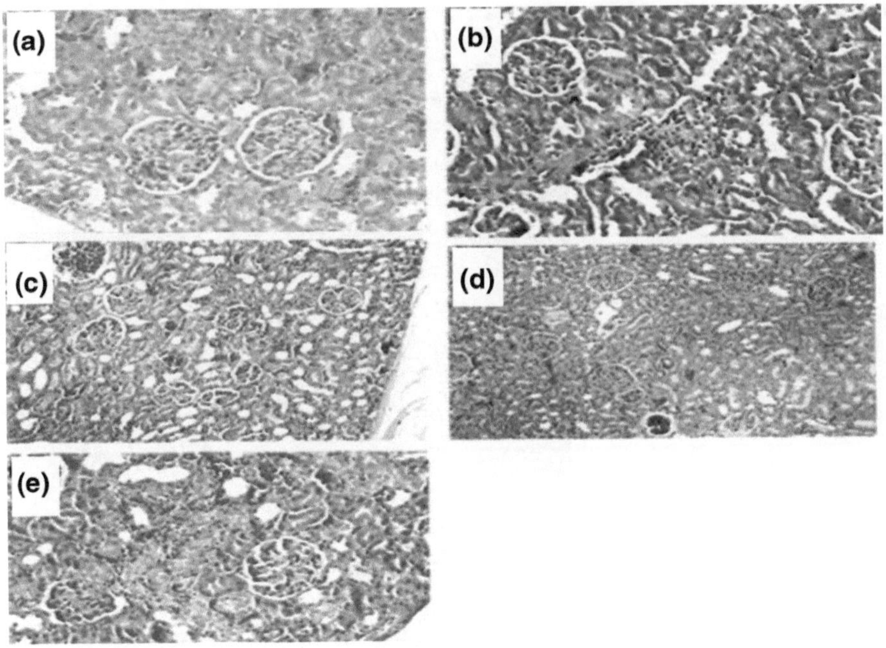

Fig. 5 Micrographs of rat's kidney histology showing: **a** normal control **b** mercuric chloride ($HgCl_2$) treated, **c** 50 mg/kg/day treated **d** 250 mg/kg/day treated, and **e** 500 mg/kg/day treated with Arogyavardini Vati (H&E × 10) [12]

of the nanoparticles. In summary, the in vitro cytotoxicity studies confirmed the nontoxic properties of the Bhasma particles. The particles showed 100 % cell viability as compared to control (medium) [11, 12].

As a second example, we present the cytotoxicity study using an Ayurvedic mercury-based formulation, the Arogyavardini Vati (see Fig. 5). This drug is a mixture of several Ayurvedic formulations such as sulfur, mercury, and asphaltum and Bhasmas of copper, iron and mica, different herbs, etc. It has been used for curing various skin and liver disorders, obesity, etc. This drug was evaluated recently in detail in terms of its toxicity that might arise from the mercury content [12].

The tests were performed on mice using doses of 50–500 mg/kg which are up to 10 times higher than the amount recommended for human beings. The study was aimed at evaluation of the safety of Arogyavardini Vati on brain, liver, and kidney in rats. Biochemical parameters, histopathology, and mercury level were assessed [12].

The optical micrographs showing the effects of chronic administration of Arogyavardini Vati on rat's kidney histology are shown in Fig. 5. The normal control, as expected, shows usual histological texture of the kidney while the mercury chloride ($HgCl_2$) treated group shows the congestion of blood vessels and epithelium disruption in proximal convoluted tubules. The histological textures of the cells after Arogyavardini Vati treatment using the doses of 50, 250, and

500 mg/kg/day do not show any damage to the kidney [12]. Different liver functions tests (LFTs) and kidney functions tests (KFTs) confirmed the notion.

It was concluded that Arogyavardini Vati in doses equivalent up to 10 times of the human dose administered to the rats under examination for several weeks does not have remarkable toxicological effects on brain, liver, and kidney of the tested animal.

Based on the discussion of this section, we can conclude that Bhasmas and similar formulations act beneficially in the bodies of living beings without posing any severe threats to normal physiological functions.

Concluding Remarks

Today, nanotechnology is emerging as a new industrial revolution worldwide and has several promises. The invasion of nanotechnology in biomedical fields has made us to dream of several opportunities, one of them being a renaissance in the field of Ayurvedic Bhasmas. There is a great chance of developing new drugs by coupling the knowledge of traditional medicine with nanotechnology. Efforts should be directed toward finding solutions to crucial issues of infectious diseases and in particular tackling with more vulnerability the come back of malaria and the devastating spread of cancer in our region.

The recent investigations on Bhasmas have undoubtedly confirmed their nanoscopic structural features. However, much has to be done to upgrade the efficiency of Bhasmas to that of modern nanodrugs. More investigations are desirable which explore the correlation between the structural nature and mechanisms associated with their therapeutic activity.

Finally, the benefit should reach in the first place to those whose traditional knowledge has basically created the foundation for the modern nanomedicine.

Acknowledgments The author would like to thank Alexander von Humboldt Foundation for supporting his several research stays in Germany which triggered his interest in Ayurvedic medicines, in particular with Bhasmas. Dr. Surendra Kumar Gautam (Department of Chemistry, Tribhuvan University, Tri-Chandra Campus, Kathmandu) is thanked for critically reading the manuscript.

References

1. Krishnamachary B, Rajendran N, Pemiah B, Krishnaswamy S, Krishnan UM, Swaminathan Sethuraman S, Sekar RK (2012) Scientific validation of the different purification steps involved in the preparation of an Indian ayurvedic medicine, Lauha Bhasma. J Ethnopharmacol 142:98–104
2. Bhatia BS, Kale PG, Daoo JV, Panchal PP (2012) Abhraka Bhasma treatment ameliorates proliferation of germinal epithelium after heat exposure in rats. Anc Sci Life 31:171–180

3. Sarkar PK, Das S, Prajapati PK (2010) Ancient concept of metal pharmacology based on ayurvedic literature. Ancient Sci Life 29:1–6
4. Jagtap CY, Prajapati P, Patgiri B, Shukla VJ (2012) Quality control parameters for Tamra (copper) Bhasma. Anc Sci Life 31:64–170
5. Wadekar MP, Rode CV, Bendale YN, Patil KR, Prabhune AA (2005) Preparation and characterization of a copper based Indian traditional drug: Tamra Bhasma. J Pharmac Biomed Anal 39:951–955
6. Unni K, Sheela KB, Vijasankar AV (2013) Scientific basis for the preparation and characterization of iron based traditional drug Annabhedi Sinduram: a materialistic approach. Int Res J Ayurveda Pharm 4:149–152
7. Mulik SK, Jha CB (2011) Physicochemical characterization of an Iron based Indian traditional medicine: Mandura Bhasma. Ancient Sci Life 31:52–57
8. Krishnamachary B, Purushothaman AK, Pemiah B, Krishnaswamy S, Krishnan UM, Sethuraman S, and Sekar RK (2013) Bhanupaka: a green process in the preparation of an Indian ayurvedic medicine, Lauha Bhasma. J Chem 951951
9. Patwardhan B, Mashelkar RA (2009) Traditional medicine-inspired approaches to drug discovery: can ayurveda show the way forward? Drug Disc Today 14:804–811
10. Chaudhary A (2011) Ayurvedic Bhasma: nanomedicine of ancient India—its global contemporary perspective. J Biomed Nanotech 7:68–69
11. Paul W, Sharma CP (2011) Blood compatibility studies of Swarna Bhasma (Gold Bhasma), an ayurvedic drug. Int J Ayurveda Res 2:14–22
12. Kumar G, Srivastava A, Sharma SK, Gupta YK (2012) Safety evaluation of an ayurvedic medicine, Arogyavardhini vati on brain, liver and kidney in rats. J Ethnopharmacol 140:151–160
13. Paul S, Chugh A (2011) Accessing the role of ayurvedic bhasmas as ethono-nanomedicine in the metal based nanomedicine patent regime. J Intell Prop Rights 16:509–515
14. Khan MS, Vishakante GD, Siddaramaiah H (2013) Gold nanoparticles: a paradigm shift in biomedical applications. Adv Colloid Interf Sci 199–200:44–58
15. Dobrovolskaia MA, McNeil SE (2007) Immunological properties of engineered nanomaterials. Nat Nanotechnol 2:469–478
16. Bhattacharya B (2011) Elucidating the nanomaterialistic basis for Ayurvedic Bhasmas using physicochemical experimentation. J Biomed Nanotech 7(2011):66–67
17. Virupaksha Gupta KL, Kumar N (2012) Characterization of Tarakeshwara Rasa: an ayurvedic herbomineral formulation. Ayurveda 33:406–411
18. Patra CR, Bhattacharya R, Mukhopadhyay D, Mukherjee P (2010) Fabrication of gold nanoparticles for targeted therapy in pancreatic cancer. Adv Drug Deliv Rev 62:346–361
19. Mohaptra S, Jha CB (2010) Physicochemical characterization of Ayurvedic bhasma (Swarna Makshika Bhasma): an approach to standardization. Int J Ayurveda 1:82–86
20. Singh N, Reddy KRC (2010) Pharmaceutical study of Lauha Bhasma. Ayurveda 31:387–390
21. Panyala NR, Peña-Méndez EM, Havel J (2009) Gold and nano-gold in medicine: overview, toxicology and perspectives. J Appl Biomed 7:75–91

Nanobiosensors: Role in Cancer Detection and Diagnosis

Andrew Gdowski, Amalendu P. Ranjan, Anindita Mukerjee and Jamboor K. Vishwanatha

Abstract The ability to detect many cancers at an early stage in its clinical course has the potential to improve patient outcomes in terms of morbidity and mortality. Nanosized components incorporated into existing clinical diagnostic and detection systems as well as novel nanobiosensors have demonstrated improved sensitivity and specificity compared with traditional cancer testing approaches. Nanoparticles, nanowires, nanotubes, and nanocantilevers are examples of four nanobiosensor systems that have been used experimentally in the context of detection and diagnosis of prostate, breast, pancreatic, lung, and brain cancers over the past few years. Nanobiosensors will begin to transition into clinically validated tests as experimental and engineering techniques advance. This paper presents examples of some such nanobiosensors for cancer diagnosis and detection.

Keywords Nanobiosensors · Cancer detection and diagnosis

Introduction

Novel nanobiosensors have the potential to vastly improve current standards and techniques for the diagnosis and treatment of cancer. Many nanobiosensor systems have proven remarkably successful in research models of various cancers and the

A. Gdowski · A. P. Ranjan · A. Mukerjee · J. K. Vishwanatha (✉)
Department of Molecular Biology and Immunology, Graduate School of Biomedical Sciences, University of North Texas Health Science Center, Fort Worth, TX 76107, USA
e-mail: jamboor.vishwanatha@unthsc.edu

A. P. Ranjan · A. Mukerjee · J. K. Vishwanatha
Institute for Cancer Research, Graduate School of Biomedical Sciences, University of North Texas Health Science Center, Fort Worth, TX 76107, USA

A. Gdowski
Texas College of Osteopathic Medicine, Graduate School of Biomedical Sciences, University of North Texas Health Science Center, Fort Worth, TX 76107, USA

R. Adhikari and S. Thapa (eds.), *Infectious Diseases and Nanomedicine I*,
Advances in Experimental Medicine and Biology 807,
DOI: 10.1007/978-81-322-1777-0_4,

future promises further translation of these nanomaterials and techniques into clinical practice. A number of advantages compared to traditional biosensor systems make nanobiosensors attractive modalities for development. Detection sensitivity can be dramatically increased due to the nano-sized components utilized in the construction of nanobiosensors. It is feasible to engineer smaller sensors that can be used to decrease lab space requirements or possibly create implantable monitors. Enlargement of the operative range is possible through extension of upper and lower limits of detection for biomarkers. Finally, fabrication costs of these nanosensors can be decreased using advanced technology and this can in turn decrease clinical healthcare testing expenditures for the detection and diagnosis of cancer.

A major reason why cancer continues to cause mortality among patients is that it is not detected early enough and current treatments for advanced stages are not as effective as treatment protocols for cancer that is caught early [1]. According to the American Cancer Society, cancer is the second leading killer across all age groups, behind cardiovascular disease, and in 2012 it is expected that 577,190 Americans will die from cancer [2]. It is promising to note that recent progress has been made in survival from cancer as evidenced from an increase in survival rate. The most current available data documenting survival from cancer during the time period from 2001 to 2007 is 67 %, which is significantly higher than the 49 % from the period of 1975–1977 [2]. The trend of increasing survival rates from all cancers has been attributed to technology that has improved treatment options and diagnostics [3]. Although the general trend is an increased cancer survival rate, there are still many types of cancer that remain quite deadly, particularly when diagnosed in a later state (pancreatic, liver, lung, etc.). Thus, one of the main issues that has limited adequate treatment of cancer is not detecting it early enough and as the cancer progresses it becomes increasingly difficult to treat [4]. Further, approximately 60 % of cancers are diagnosed after the patient's initial tumor has metastasized [3]. Developing and implementing new nanobiosensors for use in the clinical setting has the potential to vastly improve early noninvasive diagnostics, enhance imaging, and advance monitoring of treatment progress.

In order to accurately detect evidence of cancer in a biological sample, normal specimens must be differentiated from that which is cancerous. To accomplish this, various biomarkers are employed. A biomarker is a molecule that can be found in tissue specimens, blood, exhaled breath, or other fluids and can indicate a disease or health status [3]. More specifically, biomarkers can be alterations of cellular DNA, RNA, proteins, or metabolites that helps distinguish between health and disease as well as allow us to monitor treatment progress [5]. Knowledge of cancer biomarkers continues to become more sophisticated for early cancer detection and determination of prognosis. A discussion of important biomarkers currently being integrated into nanobiosensor technologies will be reviewed in later sections.

The field of biosensors has been around for nearly four decades [6] and recently much academic research has been invested in experimenting and developing new methods using nanotechnology that promise to improve current standards of disease detection. The field of nanobiosensors especially in relation to cancer

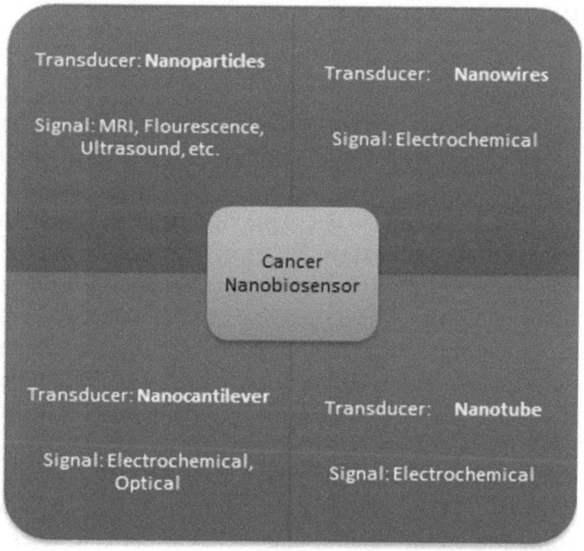

Fig. 1 Cancer biomarkers can be used in the detection and diagnosis of cancer utilizing various methods of biosensing including: nanoparticles, nanowires, nanocantilevers, and nanotubes

detection is still only in its infancy but new technological advancements and further development in understanding the complexities associated with cancer behavior should help to improve treatment outcomes by driving down many of the morbidities and mortality rates associated with cancer.

Nanobiosensors work at the level of the nanoscale broadly defined as between 1 and 500 nm or roughly 1 billionth of a meter. Development of nanobiosensors has been most successfully achieved through multidisciplinary collaborations encompassing the fields of medicine, biology, chemistry, physics, engineering, and technology. The unique contribution that nanotechnology has made are possible due to increased knowledge in the molecular biology field and the ability to exploit special properties of materials at the nanoscale through engineering and material science [7]. There are two main reasons materials have unique properties at these small sizes: (1) there is a large surface area to volume ratio which means that many of the atoms that comprise the material are a close distance from the surface of the material; (2) quantum forces are exhibited due to the fact that the material size is close to wavelengths which can excite the nanomaterial components [7].

This chapter takes the approach of first describing the characteristics known about a select number of modalities and components that comprise nanobiosensors. Particular emphasis is placed on nanoparticles, nanowires, carbon nanotubes, and cantilevers (Fig. 1). Then examples of the above-mentioned nanobiosensors are illustrated from the recent literature in the areas of diagnosis, detection and imaging of prostate, breast, pancreatic, lung, and brain cancers.

Selected Modalities

Nanoparticles: In many nanobiosensor systems, nanoparticles often play a crucial component for detecting different cancers. These particles can be made of a variety of materials and the unique properties that each possess have been engineered and exploited to achieve enhanced biomarker detection. In general, nanoparticles function in various ways to provide clinical utility whether it be through diagnostic applications such as imaging or biomarker detection, therapy application, or a combination of diagnostic and therapeutic also referred to as theranostics [8]. The first generation of nanoparticles in oncology research were simple in design and functioned as either a therapeutic cargo delivery method by allowing improved delivery of drugs to cancer sites or as an imaging agent for tumors. As research progressed in the nanotechnology field, so did the sophistication of nanoparticles. The most recent nanoparticle platforms incorporate diagnostic and imaging functions simultaneously with surface functionalized nanoparticles for targeted delivery and increased uptake at the desired area. The following sections highlight some important general categories of nanoparticles used in cancer diagnostics as well as discussions of properties that make them advantageous constituents of many nanobiosensor systems. As you will see, many of the properties of one type of nanoparticle can be applied to others.

Quantum dots: Quantum dots (QDs) are one category of nanoparticles that have been used for both in vitro and in vivo cancer detection. These are nanocrystals that are made up of semiconductor particles, consisting of an inorganic element in its core with a surrounding metal shell [9]. They generally measure less than 10 nm in diameter [10]. The benefits of using QDs in research, diagnosis, and treatment of cancer are due to some of their unique properties. The first of these properties is that their size and composition can be adjusted to give the QDs a unique fluorescence emission that can vary from 400 to 2000 nm [9, 11]. Varying wavelengths allow for tuning QDs to any color, which enables recognition and tracking of differently labeled biomarkers using only a single light source [12]. Another characteristic that makes QDs useful is their resistance to photobleaching and thus they can be used for an extended period of time [9]. One problem that is traditionally seen with imaging normal healthy tissue is that it often exhibits autofluorescence and this interferes with the signal from cancerous tissue. QDs have been engineered to have fluorescence properties in the near-infrared spectra and thus can eliminate much of this interference [9, 13]. A potential problem involved in using QDs in vivo is whether injection poses a toxic risk or not. Modifications have been made to decrease potential toxicity, however, more research needs to be carried out to determine appropriate clinical adaptability [14].

Paramagnetic nanoparticles: Contrast agents are an important consideration for detection of tumors in any region of the body. Superparamagnetic iron oxide nanoparticles (SPIONs) have been used as contrast agents in imaging cancer tumors with both magnetic resonance imaging (MRI) and computed tomography (CT) and are the most established nanomaterial in clinical practices [15]. These agents can vary in size from 50 to 100 nm for some particles to an even smaller size range of 5 to 10 nm for other particles [16]. Generally speaking, particles with smaller size tend to be taken up by a wider range of tissues including lymph nodes and bone marrow. SPIONs can also be functionalized with antibodies to various tumor markers to improve uptake and imaging procedures [17]. It has been shown through certain functionalized surface modifications, these nanoparticles can differentiate between non-apoptotic and apoptotic tumor cells. This knowledge could help guide clinicians in determining the most prudent chemotherapeutic regimen for a particular patient through easily and directly monitoring response to treatments [18]. Magnets located external to the body surface can be used to exploit intrinsic magnetic properties of these nanoparticles by directing them to a tumor site. This targeting mechanism can be coupled with cytotoxic drugs encapsulated within the nanoparticle to greatly increase the concentration of drugs at tumor sites. The goal of this therapy is to decrease the total amount of drug being administered which in turn minimizes many potential toxic side effects patients must tolerate while being able to monitor this process through MRI [19].

Nanoshells: Nanoshells comprise another category of nanoscale particles that combines a dielectric core with a surrounding metal shell often composed of gold. These particles are particularly useful due to their ability to be tuned to different optical resonances by adjusting the size of the core and shell similar to the QDs mentioned earlier [20]. Nanoshells extend their effectiveness by incorporating the ability to be used as an imaging component in addition to having the ability to load drugs into the core [21]. Antibodies have also been conjugated to nanoshells allowing for targeted delivery of its cargo to specific tumor types [22]. One common means of increasing circulation half-life of nanoshells and other nanoparticles is coating the outer surface of the nanoshell with poly ethylene glycol (PEG) otherwise known as PEGylation. In addition to increasing in vivo circulation, PEGylation has also been shown to decrease toxicity of gold nanoshells, decrease aggregation, improve avoidance of macrophages, and the reticuloendothelial system [23]. Another interesting application of nanoshells that has been reported is the ability to generate heat after exposure to near-infrared light [24]. Heat generated at the tumor site where nanoshell have aggregated can cause a photothermal ablation effect to kill cancer cells [25]. In this way nanoshells can have a diagnostic imaging function combined with cytotoxic consequences.

Gold nanoparticles: Another class of particles which have shown promise in diagnostics is nanoparticles made from colloidal gold. Gold nanoparticles mostly exist as gold nanospheres, which exhibit an intense ruby color in aqueous solutions [26]. The fascinating optical properties of gold nanoparticles arise from localized surface plasmon resonance (LSPR), in which valence electrons in gold nanoparticles oscillate coherently with incident light at specific frequency [27]. Part of the energy absorbed by gold nanoparticles is emitted in the form of scattered light, which forms the basis of gold nanoparticle-based optical imaging [28]. The rest of the energy decays in a nonradiative form which is converted into heat, which can be used in killing cancerous cells and hence play a role as photothermal therapy. Gold nanoparticles may find use in multiple diagnostic potential areas such as labeling precancerous cervical biopsies [29] as well as other prospective uses in cancer treatment [30].

Liposomes: Liposomes are nanoparticles composed of a lipid bilayer which can be used in transporting imaging agents or drugs to the site of a tumor. Along with other types of nanoparticles, liposomes use the property of enhanced permeability and retention (EPR) to enter tumor vasculature and remain there. The phenomenon of EPR stems from a structural understanding of vasculature associated with solid tumors. Nanoparticles have increased ability to extravasate through less tightly formed endothelial junctions. Once in the tumor microenvironment, increased retention is due to inadequate lymphatic drainage [31]. In order to utilize the EPR effect, liposomes are usually engineered with a size of less then 200 nm to improve accumulation in the tumor [32]. Liposomes can also be labeled with a targeting component in order to increase cellular uptake at target tissue.

Nanowires: Nanowires also make up an interesting category of biosensors. A nanowire can be defined as a material consisting of millimeters in length but achieves a diameter that is measured in the nanometer range. Nanowires composed of silicon have been best characterized [33] due to advancement of manufacturing techniques at the large scale and represent an example that may prove useful as point of care diagnostic devices [34]. Their ability to detect biomarkers in a sample, such as a drop of blood, rely on the nanowire being a field effect transistor (FET) device. The basic concept behind biosensing in this application is when a biomolecule of interest comes into contact with the nanowire, a measurable change in the electric field of the nanowire occurs due to increased resistance. Further, direct detection utilizing nanowires has been show to be extremely sensitive for a number of different substances including: metal ions, nucleic acids, proteins, protein-DNA interactions, small molecule–protein interactions, cells, and viruses by surface functionalization techniques [34]. Specific applications will be mentioned in the following sections regarding how nanowires can be used to detect cancer biomarkers.

Carbon nanotubes: There has been much interest in the area of carbon nanotubes (CNTs) since their discovery over 20 years ago by Japanese physicist Iijima [35]. Carbon nanotubes are structures composed of either single-walled or double-walled carbon molecules self-arranged in hexagon pattern with diameters ranging from 0.3 to 100 nm [36]. There are several characteristics that make CNTs attractive for use in nanobiosensors. High heat resistance and conductivity allow CNTs to remain operable even while exposed to high temperature or electrical currents. CNTs also exhibit a high degree of tensile strength that permits them to resist permanent deformation after exertion from physical forces. Often CNTs are placed into two general categories based upon how they sense biomarkers, they can generate an electrochemical signal based on oxidation–reduction reactions that occur or as FET-based detection which derives its signal from charges generated on the carbon nanotube surface [37] similar to nanowires described above. Nanotubes of both single and multiple-walled types have been used in an application known as nanotube forests. This technique requires nanotubes to be lined up next to each other in a parallel fashion and then antibodies for specific biomarkers are attached to the surface. After the binding interaction of antigen and antibody, an additional binding event can occur between an antibody coupled to a signaling agent which binds the other side of the antigen and essentially makes a sandwich. Measurement of the amount of antibody binding events allows for quantification of cancer biomarkers. The other common method of detection incorporating carbon nanotubes is the previously mentioned FET method, in which a nanotube is connected to an electrode at either end using photolithography and functionalized to be able to bind biomarkers. Once these biomarkers are bound, a decrease in conductive properties of the nanotube is detected that is proportional to the amount of binding that has occurred.

Nanocantilevers: The last general category of nanobiosensors that will be discussed in this paper is nanocantilever devices. Cantilevers are able to conduct biosensing through the principle that these tiny probes naturally vibrate at a certain frequency dictated by mechanical and mass properties. When a biological molecule binds to this nanoscale probe it will alter baseline probe frequency, this change is typically measured by a difference in the characteristics of the light deflection pattern of the probe or through electrical means [38]. Different nanocantilevers systems can detect different phenomenon regarding behavior of the probe. Two common types of probes exist: static, which when bound by a biological molecule will bend to one side or another, and resonant probes, which when bound by a biological molecule change their resonance frequency. Interestingly, it has been observed that cantilevers with diameter in the micrometer or thicker range tend to decrease resonant frequency when attaching mass while nanosize cantilevers tend to increase frequency when mass such as antibodies are attached. This is explained by effecting a property called the net stiffness constant of the nanocatilever as mass is added [38]. Nanocantilevers have been described as a simple replacement to PCR reactions and detection methods because they do not require costs associated with sample preparation such as time and expensive materials. Nanocantilevers could also be used to monitor various cancer biomarkers and expand the usefulness of microarray technologies [39].

Applications in Various Cancers

Prostate Cancer

The PSA screening test is the most widely used screening tool for prostate cancer and the long controversy surrounding it illustrates the need for more sensitive and specific tests that can better determine a patient's course of treatment [40]. Prostate-specific antigen (PSA) is a glycoprotein secreted by epithelial cells of the prostate gland that was first discovered and utilized for forensic purposes [41]. In 1994, the Food and Drug Administration approved the PSA test for prostate cancer screening in men over 50 along with digital rectal examination and this was the standard screening protocol for years [42]. Most recently, in May 2012, the U.S. Preventative Services Task Force (USPSTF) recommended against the use of the PSA screening test [43]. The USPSTF's report highlights the need for a better diagnostic test that can distinguish between aggressive cancer and types of cancer that are nonprogressive or progressing so slowly that it won't affect the patient's life [43]. Current work in nanobiosensors and tumor marker discovery may help to create and validate such a test that can spare patients unnecessary procedures and biopsies if they are unlikely to benefit.

Nanoparticles: Although the PSA test has been scrutinized for its clinical utility as a general screening test, it has been accepted as an important tool in monitoring the status of cancer recurrence after radical prostatectomy as well as determining response to treatments. One emerging nanotechnology assay technique that may prove useful clinically for PSA testing is bio-barcodes [44]. These bio-barcodes have been shown to be hundreds of times more sensitive than conventional assays used commercially [45]. The concept behind bio-barcodes is that attached to gold nanoparticles are antibodies to PSA and hundreds of strands of DNA barcodes. Magnetic bead particles also have PSA antibodies attached. When PSA antibodies of the magnetic beads and gold nanoparticles encounter the PSA glycoprotein they form a complex by sandwiching the PSA molecule between them. A magnetic field is then used to separate the complex from the sample. Subsequently, the amount of DNA barcode that has been separated can be quantified to determine the amount of PSA in the sample [45, 46].

Detection of cancer cells in blood can also serve as an indicator of a patient's cancer status. Prostate cancer cells can be tagged with gold nanoparticles so that the cells can be detected in blood through a photoacoustic flowmeter. Through this method cancer cells can be detected, making it have great potential as a serum detection method [47]. Another important detection modality that utilizes gold nanoparticles is to detect whether prostate cancer cells have migrated to lymph nodes. In order to determine if migration has occurred, highly lymphotropic superparamagnetic nanoparticles have been shown as effective clinical imaging agents to detect small metastasis in lymph nodes through MRI [48].

Nanoparticles that fall into the category of theranostic nanoparticles may play a future role in prostate cancer treatment. These multifunctional nanoparticles allows for simultaneous administration of therapeutic and diagnostic agents and can even be taken a step further by having a targeting agent attached to the outer portion. This addition helps to improve nanoparticle cargo uptake at target cancer cells [49]. These nanobiosensors hold promise for improved detection and survival for patients with prostate cancer, whether it be on initial diagnosis, monitoring, or planning the most effective treatments.

Nanowires: Another category of nanobiosensors that may prove beneficial for prostate cancer patients is nanowires. Silicon nanowires have been reported to be able to detect PSA at the level of 1 fg/ml of PSA through optimization of dimensions of the nanowire as well as amount of doping concentrations for antibody functionalization during nanosensor construction [50]. Further, multiplexed tumor antigen recognition is possible utilizing silicon nanowires. Zheng et al. were able to construct arrays composed of nanowires functionalized with antibodies against PSA, PSA-alpha1-antichymotrypsin, carcinoembryonic antigen, and mucin-1. They were able to reliably detect these biomarkers at femtomolar concentrations in undiluted serum samples [51]. Another system reported by Stern et al. uses a two-step approach that incorporates microfluidic purification chips which capture multiple biomarkers from whole blood, concentrates the analyte with the biomarkers of interest, and releases the biomarkers for quantitative detection with silicon nanoribbon detectors. This technique reduces the minimum required sensitivity of the system [52].

Carbon Nanotubes: Carbon nanotubes forests have been shown to present advantages for cancer antigen detection. One group reported up to 15-fold increased detection sensitivity of cancer antigens based on their ability to densely pack and immobilize more surface antibodies compared with sensors utilizing a flat surface [53]. The proteins: PSA, prostate-specific membrane antigen, platelet factor-4, and interleukin-6, all known to be increased in prostate cancer patient serum, were measured simultaneously through an array based on carbon nanotube forests [54]. Examples of carbon nanotube forests are increasing in the literature and represent important assays that can detect multiple antigens during one measurement with important implications for better identification and improved risk stratification for cancer patients.

Nanocantilvers: PSA antigen antibody binding has also been detected using piezoeclectric nanomechanical cantilevers. Analysis of one type of cantilever used in PSA detection showed resonance frequency change as a result of surface stress placed on the nanocantiliver rather than from increase in protein mass associated with binding events [55]. Some reported benefits of using nanocantilever systems in cancer detection are there are no requirements for fluorescent or radioactive labeling, detection can take place in liquid samples, and this technology can easily be translated to lab on a chip techniques providing point of care diagnostics [56].

Breast Cancer

Although there have been many advances in the detection and treatment of breast cancer, it still remains the most prevalent cancer in women with about 230,480 women being diagnosed with breast cancer in 2011, and almost 40,000 deaths [57]. Current strategies for treating and diagnosing women with breast cancer often carry noteworthy side effects, require invasive procedures and leave women permanently scarred [58]. The application of nanobiosensors has been applied to multiple facets of breast cancer diagnosis. The areas that will be discussed in the following paragraphs are early detection methods, improved prognosis estimation, advances in sentinel lymph node biopsies, screening, and monitoring treatment responses in breast cancer.

One biomarker that has provided both diagnostic and therapeutic usefulness in breast cancer is HER2, which is a plasma membrane-bound receptor tyrosine kinase encoded by the ERBB2 gene [59]. Amplification of this gene occurs in about 30 % of breast cancer patients and when detected indicates a worse prognosis [17].

Nanoparticles: There have been a numerous examples of nanoparticle systems utilizing HER2 introduced for detection of breast cancer. Wu et al. showed that attachment of IgG and streptavidin to QDs can be used to fluorescently label HER2 positive cells on both histologically fixed cells as well as live cells. The benefit of using QDs over traditional staining procedures is that with QDs the authors were able to show a stronger targeted fluorescent signal that was more stable than traditional techniques. In addition, this form of labeling can be used for multiple cellular targets and simultaneously excite various colored QDs with only one light source [60]. Similarly, a study done by Chen et al. showed that their QDs-based probe was more accurate and sensitive than immunohistochemical techniques in detecting HER2 in clinical breast cancer samples [61]. Other fluorescently labeled silica nanoparticles have also shown a high degree of sensitivity to breast cancer cells when conjugating anti-HER2/neu to the nanoparticle [62].

Multiple examples in the literature demonstrate the use of superparamagnetic nanoparticles conjugated to various targeting moieties to serve as targeted contrast agents in MR imaging. The HER2/neu receptor has been used as a target for streptavidin conjugated superparamagnetic nanoparticles made from iron oxide. The nanoparticles in this instance were imaged using MR molecular imaging and proved to have proportional contrast with the amount of HER/neu expression on the surface of the cells that were being studied [17]. LHRH conjugated to SPIONs have shown utility through a mouse model for in vivo detection of lymph node metastasis along with MRI [63]. In addition, SPIONs have been functionalized with a recombinant peptide that targets urokinase-type plasminogen activator (uPA) receptor known to be overexpressed in breast cancer tissues [64].

Herceptin, the antibody that targets HER2, has been attached to gold nanoparticles so that the nanoparticles could be used as a targeted contrast agent while using the imaging method of optoacoustic tomography. The authors of this study were able to show that using this method in a gelatin breast model, it was possible to sensitively detect small tumors that were implanted up to 6 cm deep [65]. Herceptin conjugated to iron oxide nanoparticles was also tested using ultrasound as the imaging modality and was found to be able to elicit a significant ultrasound signal and thus has the potential to be used in the future as an inexpensive detection method for breast cancer [66].

Another strategy incorporates the use of ultrasound for both diagnostic purposes and therapeutic delivery of drugs. Two different nanoparticles were used in this study: polymeric micelles loaded with doxorubicin and echogenic nanobubbles loaded with doxorubicin in the walls of the nanobubble. The polymeric micelles use the well-known EPR effect that is seen in all tumor vasculature for delivery to the tumor site. The nanobubble also utilize the EPR and once at the tumor site the nanobubbles coalesce into a microbubble. This allows for increased drug delivery from the nanobubble, increased drug delivery from the corresponding micelle and also a high quality contrast agent for ultrasound imaging. In addition, the ultrasound can be adjusted to more precisely control where the drug in the nanobubbles is released [67].

Gold nanoparticles have been functionalized with a variety of surface targeting molecules such as transferrin. The transferrin receptor is important for iron uptake in normal cells and cancers. Tumor cells are in a stage of rapid proliferation so they up regulate transferrin receptors to maintain growth [68]. It has been shown that transferrin-mediated cellular uptake of gold nanoparticles is more pronounced by a factor of 6 in breast cancer compared to non-cancerous cells [69].

Another class of nanoparticles that uses HER2 as a target are nanoshells. Nanoshells are particles that are made of a dielectric core and covered with a gold shell. These particles can be adjusted similar to QD to have different optical imaging properties and absorb near-infrared (NIR) light [70]. Nanoshells have added benefit because they have also been shown to have dual therapeutic and diagnostic imaging properties. They are able to reflect light and thus be used in imaging as well as absorb light allowing the nanoshells that have been taken up in the cells to increase in thermal activity thereby killing breast cancer cells [22].

In addition to the potential use of nanoparticles for detection of primary breast cancers that have remained localized, nanoparticles have also shown promise in assisting surgeons in identifying breast cancer that has metastasized to the lymph nodes. One challenge that surgeons face when performing a masectomy after detection of breast cancer is determining the extent, if any, the primary cancer has spread to lymph nodes [71]. Sentinel lymph node mapping has greatly improved surgical treatment of women with breast cancer due to less nodes having to be removed for examination by the surgical pathologist which in turn results in a more concentrated histological examination and decreased morbidity [72]. The current clinical standard generally involves the surgeon taking one of the two approaches, either injecting vital blue dyes or using radioactive technetium-99 m sulfur colloid [73]. Both ways present drawbacks due to radioactive exposure to patients or limited visibility using dyes. QDs and ICG-human serum albumin nanoparticles have been proposed as mitigating the above problems and providing improved intraoperative imaging of sentinel lymph nodes using fluorescence imaging systems that utilize NIR wavelengths and improved contrast compared to current approaches [74, 75]. A study by Ballou et al. has shown that PEG coated QDs could be used as a method for identifying sentinel lymph nodes in mice [76]. Again, in other studies, single molecule targeting is used and sentinel lymph nodes are detected using QDs [77, 78].

Sentinel lymph node mapping can also be done using gold nanotubes. The benefits of one reported method is that it uses photoacoustic imaging so it is noninvasive and can be used to sensitively detect sentinel lymph nodes in tissues that are up to 33 cm deep [79].

Another interesting area that is currently being researched is in vivo characterization of tumor subtypes so that specific treatment protocols can be rapidly implemented and carried out. Again, QDs have been used to study and differentiate cancer subcategories by using different markers in a single tissue specimen [80]. In addition, dendrimer nanoparticles have been created which can be dually utilized for imaging using MRI or NIR fluorescent modalities in the single probe. This has been shown effective in mouse sentinel lymph node mapping as well [81, 82].

One common way of obtaining a biopsy of a breast lump is through fine needle aspiration (FNA). Issues can arise using FNA as a diagnostic test for cancer due to the high amount of false negatives and false positives originating at least partially from the limited biopsy sample of this method. Nanobiosensor techniques are ideal for the specific difficulties of analyzing the FNA sample. Lee et al. devised a method that can label breast cancer cells with magnetic nanoparticles that have been obtained through a fine needle aspiration. They then used a diagnostic magnetic resonance probe to detect characteristics often seen in cancer cells as compared to normal non-cancerous cells. This method illustrates the ability to maintain a sensitive test while only requiring a small amount of tissue from the fine needle aspirate [83].

Nanowires: One of the capabilities of nanowires is to detect protein-DNA interactions. An illustration of the importance of using this interaction for breast cancer detection can be seen with an example from the estrogen receptor (ER). The ER plays an important role in many normal physiological processes but an abnormal ERalpha exists in over half of breast cancers where it is a ligand-dependent transcription factor and plays a role in cancer initiation and progression [84]. A sensitive assay for detection of ERalpha interaction with double-stranded (ds)DNA has been developed. In this assay a silicon nanowire was surface functionalized with dsDNA of wild-type estrogen receptor elements. When nuclear extracts of breast cancer cells expressing the ERalpha receptor were evaluated using the nanowire apparatus there was notable detection due to abnormal receptor binding to dsDNA functionalized nanowire and this was distinguishable from normal ER binding. The conductance change reported for this nanowire was able to detect a biomarker protein concentration as low as 10 fM [85].

Carbon nanotubes: The use of carbon nanotubes has been demonstrated in a number of different purposes for detection of breast cancer. Using carbon nanotubes functionalized with IGF1R and Her2 antibodies (known to be upregulated in some breast cancer cells) one group created a device able to detect single circulating cancer cells from human breast cancer cell lines MCF7 and BT474 in two microliter drops of blood [86]. A group used paclitaxel (PTX) as a therapeutic agent and coupled it via a cleavable ester bond to the PEGylated nanotube surface. This construct was tested in a murine 4T1 breast cancer model and the results depicted a 10-fold increase in tumor accumulation than PTX alone followed by improved tumor suppression compared with clinically used Taxol [87]. They also reported the coupling of other agents like Pt(IV) prodrug PEGylated carbon nanotubes to improve the pharmacokinetics and therapeutic effects [88].

Nanocantilevers: Detection of BRCA1 mutations can be an important tool in determining risk for development of breast cancer, ovarian cancer, and prostate cancer. A cantilever assay has been developed that can detect a single nucleotide polymorphism (SNP) for the BRCA1 gene. The benefits of this assay is that it is highly specific and it only takes about 30 min to complete the process of DNA immobilization of the target DNA of interest, hybridization, washing, and readout [89].

Pancreatic Cancer

In 2012, approximately 37,390 people in the United States died from pancreatic cancer and about 43,920 new diagnoses were made [2]. Oftentimes cancer of the pancreas develops without early warning signs and symptoms and this factors into the reason that pancreatic cancer has such a low survival rate. The survival rate for pancreatic cancer has unfortunately remained somewhat stagnant over the past 30 years with an overall 5-year-survival rate of less than 5 % [90]. The majority of people who are diagnosed with pancreatic cancer will eventually die, either from metastasis to distant organs or from local tumor invasive of the superior mesenteric artery or celiac artery, however in either case the cancer is inoperable when diagnosed at this stage [91]. Thus, when surgical intervention is appropriate the patient will have the highest chances of survival and detection and diagnostic approaches should be focused on the early stages of pancreatic cancer development. Nanobiosensors approaches offer the capability to sense small quantities of biomarkers in small sample volumes as well as improve imaging techniques and potentially advance early detection of pancreatic cancer.

Nanoparticles: There are a number of different categories of nanoparticles that have been studied in pancreatic cancer systems. SPIONs represent one of these groups that have been used in multiple experimental applications for identification and treatment of pancreatic cancer. Different targeting ligands have been utilized in successful surface functionalization for targeting of these iron oxide nanoparticles including [92]: urokinase plasminogen activator receptor (uPAR) which is a surface receptor that has increased expression on pancreatic tumor cells and stromal cells that surround the tumor [93], as well as folate. These multifunctional nanoparticles have potential for use in diagnostics, with both MR imaging and fluorescent imaging after a dye is loaded [94]. Yang et al. demonstrated that single-chained epidermal growth factor receptor antibody to the surface of either QDs or magnetic iron oxide nanoparticles results in targeted delivery to pancreatic cell and could be beneficial in drug and imaging delivery [95]. ZnO QDs have also been used as a signaling agent in sandwich immunoassays to detect CA19-9 cancer biomarkers with detection methods using both square wave stripping voltammetry and photoluminescence [96].

Another example of nanoparticles that has been used to target pancreatic cancer cells is organically modified silica nanoparticles measuring approximately 20 nm in diameter. These nanoparticles were surface functionalized with transferrin, anti-claudin 4, and antimesothelin and used the fluorophore rhodamine B as an imaging agent. This technique was found to be simple and suitable for optical bioimaging studies [97]. Similarly, mesoporous silica nanospheres where functionalized in a different study with a targeting ligand and a covalently linked Gd(III) through a disulfide moiety. It was demonstrated to be an effective MRI contrast agents in the in vitro pancreatic cancer system but when tested in vivo, the disulfide was quickly cleaved which highlighted the challenges associated with release kinetics of linker systems [98].

As mentioned earlier, there has been concern about toxicity associated with injecting humans with semiconductor metals used in construction of QDs. In response to this trepidation, over recent years, a trend has been set to engineer non-cadmium-based QD that avoid the potential barrier to widespread adoption into clinical practice. One such effort engineered QDs that were composed of an indium phosphide core and a shell made of zinc sulfide that were able to target live pancreatic cancer cells after functionalization with anticlaudin-4 or anti prostate stem cell antigen [99]. Another study took a different approach in attempting to reduce potential toxicity of CdTe/ZnS QDs by effectively encapsulating them into a triblock polymeric nanomicelle. This was effective at showing decreased toxicity while at the same time effectively accumulating in pancreatic tumors. The polymeric micelles were then conjugated with anti-mesothelin antibodies for targeting to pancreatic tumors and the size of the micelle and enclosed QD were measured to be around 120 nm [100]. Zaman et al. reported effective functionalization of QDs with a single domain antibody (2A3) which targets CEACAM6. This could be used as a new biomarker for pancreatic cancer. The authors also report that using single domain antibodies is better for targeting because they are most stable, are less likely to cause nanoparticles to aggregate, and are more cost-effective than using a traditional antibody attachment [101]. Various techniques [102] and materials have been reported that incorporate numerous targeting ligands and core materials being effective in directing QDs to pancreatic cancer for detection and diagnosis [103–107].

In addition to the other nanoparticles listed, gold and silver nanoparticles have been successfully employed for detection and diagnosis of pancreatic cancer [108, 109]. In a specific example gold nanoparticles were used to label human pancreatic cancer tissue. The authors reported that covalently attached F19 human monoclonal antibodies to gold nanoparticles were effective at labeling pancreatic carcinoma tissue and were able to visualize cancerous areas through darkfield microscopy [110]. In a different approach, functionalized gold nanorods and silver nanoparticles proved their capacity to be used as probes in pancreatic cancer detection employing darkfield microscopy or TEM [111].

Most of the experiments that use nanoparticles to view MRI images discussed so far utilize the strategy of labeling cancerous cells by targeting a receptor that has been upregulated on the surface of cancerous cells compared to normal healthy cells. An alternative approach to labeling has been proposed by a group that has created a method for labeling normal cells in pancreatic ductal adenocarcinoma. They created nanoparticles to target bombesin receptors which are known to reside on healthy cells of the pancreas. In this approach they were able to improve T2-weighted pancreatic images on magnetic resonance imaging [112].

Carbon nanotubes: Carbohydrate antigen 19-9, an important tumor antigen for pancreatic, gastric, colorectal, and hepatic cancers, has been used in the construction of a novel immunosensor incorporating carbon nanotubes, gold nanoparticles, and SiO_2 nanoparticles as components. The process used to create this nanosensor was the peripheral surface of the CNT was covered with bovine serum albumin (BSA) molecules, gold nanoparticles were attached to the BSA-CNT, next electrochemical deposition of gold was added with the first gold nanoparticles providing nucleation sites for this composite layer. These initial steps provided a large surface in which CA19-9 antibody could be immobilized to and serve as a sensing component. For signal improvement and thus better sensitivity of the assay, Si02 nanoparticles were used to decorate the secondary antibody used in this sandwich immunoassay. Based upon experiments incorporating various concentrations of CA19-9 the authors demonstrated a detection limit 100 times lower than current ELISA standards used in clinical practice [113].

Multiwalled carbon nanotubes have also be used to construct a genetic fingerprint map for pancreatic cancer. In one study, the authors combined random amplified polymorphic DNA and multiwalled carbon nanotube electrochemical sensing to detect differences of guanine and deoxyguanine triphosphate in peripheral blood DNA samples of pancreatic cancer patients and controls [114].

Lung Cancer

Lung and bronchus cancer is estimated to account for nearly 226,160 new cases of cancer diagnosed in the United States. It is second among new cancer diagnosed in men and women every year behind prostate and breast cancer. However, lung and bronchus cancer account for more than three times the amount of deaths than prostate cancer and nearly twice as many deaths in women from breast cancer [2]. This makes lung cancer truly the most deadly of all cancers. The same importance of early diagnosis that has been stressed in the other cancers should be reinforced for this lethal form of cancer. Current methods of diagnosing and classifying lung cancer are imaging and biopsy-based, however these procedures are often invasive and require large imaging equipment.

An important component in lung cancer diagnosis and treatment options is differentiating various types and subtypes of lung cancer as well as stage due to the essential role this knowledge plays in selecting treatments [115]. Adenocarcinoma, squamous cell carcinoma, large cell carcinoma, and small cell carcinoma are categories of lung tumors that have been distinguished. These categories exhibit behavior that responds differently to various treatment avenues. Thus, earlier and more accurate diagnosis of lung cancer utilizing nanobiosensor technology can play a tremendous role in the outcome of patients.

Nanoparticles: Efforts are being pursued to create valid clinical screening tests for lung cancer and other cancers based on chemical patterns recognized in exhaled breath. Barash et al. proposed a method for detection of various volatile organic compounds (VOCs) with gold nanoparticles. VOCs were collected from the headspace of numerous lung cancer cell lines. The researchers were able to create a chemical profile and determine which type and subtype of cancer cell the VOCs came from based on detection of binding gold nanoparticles (GNPs) with VOCs ligands [116]. In another study that also used a functionalized GNP-based array linked with gas chromatography–mass spectroscopy (GC–MS), it was shown that cancer patients with not only lung, but also breast, prostate, and colorectal cancers, were able to be categorized based on their breath samples [117]. While it has long been established that VOCs could differentiate lung cancer patients from healthy patients through GC–MS [118] there are drawbacks to using mass spec techniques compared to proposed nanotechnology analysis such as additional expensive equipment, time requirements of sample preparation procedures, and the need for knowledgeable technician to manage the GC–MS whereas nanobiosensors eliminate or drastically reduce some of these challenges (Table 1).

Nanowires: Similar to the discussions of other cancers mentioned, nanowires have also been used in the experimental context to capture and analyze circulating lung cancer cells. One group fabricated a quartz nanowire array and functionalized it with antibodies against epithelial cell adhesion molecules (EpCAM) which are upregulated on the surface of lung carcinoma cells. Blood samples were fed through a microfluidic cell capture apparatus and the cells were captured through binding the nanowire and then imaged with laser scanning cytometry. In this method, circulating tumor cells that only constitute a few cells per milliliter of whole blood could be detected [33].

Nanowires have also been utilized to detect two other potential biomarkers for lung cancer: IL-10 and osteopontin (OPN). A silica nanowire was used with a capture antibody that could bind either IL-10 or OPN. After binding cancer antigen to antibody, an additional detector antibody attached to an alkaline phosphatase also binds to the antigen. The second binding triggers the alkaline phosphatase to dephosphorylate the p-nitrophenyl phosphate substrate. When the reaction was measured the anodic peak current had a direct linear relationship with the concentration of the biomarkers being studied [119].

Carbon nanotubes: One example of incorporating carbon nanotubes into lung cancer detection is shown with a sensor array of single-walled carbon nanotubes that has been shown competent to distinguish breath profiles utilizing VOCs of lung cancer patients from healthy subjects [120].

Brain Cancer

Primary and metastatic brain tumors have the potential to cause devastating consequences. While brain tumors are not the most common cancer to cause death in the adult population, they represent the most common cause of cancer associated death in the pediatric population [121]. Consistent with the theme of every other cancer discussed to this point, early detection of brain cancers is associated with improved prognosis and chances of survival. Further, two challenges specific to cancers originating in the brain complicate the ability of diagnosis and treatment: (1) tight junctions of the capillary endothelium that make up the blood-brain barrier (BBB) and (2) delicate structures may be in close contact to tumor consequently instigating severe morbidity or mortality if damaged.

Table 1 Examples of various cancer biomarkers/targets and modalities as well as the biosensor principles that have been used for detection

Cancer	Biomarker/target	Modality	Biosensor principle	References
Prostate	PSA	NPs	Bio-barcodes	[44–46]
	PSA	NPs	Photoacoutic	[47]
	PSA	NPs	MRI	[48]
	PSA	Nanowires	FET	[50]
	PSA, PSA-alpha 1, Antichymotrypsin, CEA, Mucin	Nanowires	FET	[51]
	PSA	Nanoribbons	FET	[52]
	PSA, Platelet factor-4, IL-6	Nanotubes	Electrochemical	[54]
	PSA	Nanocantilever	Mechanical resonance	[55, 56]
Breast	CA15-3	Nanoribbons	FET	[52]
	HER2	NPs	Fluorescence	[22, 60–62, 70, 77]
	HER2	NPs	MR	[17, 83]
	LHRH	NPs	MR	[63]
	uPA	NPs	MR/fluorescence	[64]
	HER2	NPs	Optoacoustic	[65]
	HER2	NPs	Ultrasound	[66, 67]
	Transferrin	NPs	Fluorescence	[69]
	Lymph node	NPs	Fluorescence	[75, 76, 78]
	Lymph node	Gold nanotube	Photoacoustic	[79]
	CD44v6 and CD24	NPs	Fluorescence	[80]
	Lymph node	NPs	MRI/fluorescence	[22, 81]
	Eralpha	Nanowires	FET	[85]
	IGFIR and HER2	Nanotubes	FET	[86]
	BRCA1	Nanocantilever	Microelectronic	[89]
Pancreatic	uPAR	NPs	MRI/fluorescence	[93]
	EGFR	NPs	MRI/fluorescence	[95]
	CA19-9	NPs	Electrochemical/ fluorescence	[96]
	Transferrin, claudin 4, mesothelin	NPs	Fluorescence	[103, 105]
	Anastimide	NPs	MRI	[98]
	CEACAM16	NPs	Fluorescence	[10]
	F19	NPs	Darkfield microscopy	[110]
	Bombesin	NPs	MRI	[112]
Lung	Lymph node	NPs	Fluorescence	[74]
	VOCs	NPs	GC–MS	[116, 117]
	EpCAM	Nanowires	Cytometry	[33]
	IL-10 and OPN	Nanowires	Electrochemical	[119]
	VOCs	Nanotubes	GC–MS	[120]
Brain	F3 peptide	NPs	MRI	[124]
	GPNMB	NPs	MRI	[125]
	Chlorotoxin	NPs	MRI	[127]

Nanoparticles: The brain offers a particularly promising area for nanoparticles to improve not only delivery of therapeutics but also imaging agents for better diagnosis and detection standards. The blood-brain barrier (BBB) is notable for allowing only drugs and small molecules with a molecular mass below 400–500 Daltons and high lipid solubility to cross through into the brain parenchyma [122]. These restrictions can be overcome through engineering various nanoparticles that bypass requirements of the BBB.

Many traditional imaging technologies such as positron emission computed tomography (PET), single photon emission computed tomography (SPECT), CT, X-ray, and MRI used clinically can benefit from enhancing tumor detection through nanoparticle-based modalities. The field of theranostics is working to capitalize combining diagnostic and therapeutic applications into a single preparation often utilizing current clinically available imaging systems as well as newer optical imaging. Both organic NPs as well as magnetic NPs have been used for theranostic applications in brain cancer with magnetic NPs being the most important so far due to their intrinsic MRI function [123].

One example that uses organic nanoparticles for theranostic application was shown as a proof of concept experiment using polyacrylamide nanoparticles. The NPs were functionalized with F3 peptide to target the NPs to glioma tumor vasculature. Encapsulated within the polymeric NP were iron oxide nanoparticles and a photosensitizing agent. The iron oxide is responsible for the MRI imaging element of this multifunctional nanoparticle and when the photosensitizing agent is activated by light irradiation, it generates a singlet oxygen cytotoxic component. In this system they were able to both treat and image rat gliomas in a targeted manner [124].

Further, use of functionalized magnetic NPs has been carrier out in interesting theranostic applications. In one example, a focused ultrasound was used to increase the permeability of the BBB in a locally directed manner, next magnetic targeting was applied to direct magnetic nanoparticle to the target site. This MRI monitored process led to an increased dose of cytotoxic medication that was loaded into the NP being delivered to the target [132]. Another group has taken a surface functionalization approach to magnetic nanoparticles. Glycoprotein non-metastatic melanoma protein B (GPNMB) is overexpressed by glioblastoma cells and an antibody against this target was conjugated to paclitaxel-loaded magnetic nanoparticle. This magnetic nanoparticle system was shown to effectively deliver the cytotoxic agent while being able to monitor treatment through MRI in a rat model [125]. Other examples of technologies based on magnetic nanoparticles for detection of brain cancer can be found in the literature [126, 127].

In addition, newer optical-based imaging methodologies incorporating fluorescent nanoparticles, bioluminescence, and optoacoustic tomography provide exciting possibilities. The radiation exposure from light is non-ionizing and can be used frequently without the same worry as X-ray or CT-based cumulative exposure [128]. Optical approaches have also been used to target tumors for imaging such that a near-infrared dye conjugated to a tumor-specific small peptide can show improved retention within tumor tissue being detected by a continuous-wave optical imaging system when using the rat tumor model CA20948 [129].

Nanowires: One intriguing demonstration in the application of nanowires comes from a collaboration of researchers that developed a system to use platinum nanowires with a 0.6 μm diameter to travel into the spinal cord vascular bed and record electrical stimulation. These researchers are working on improving their technique and envision replacing the platinum nanowire with a biodegradable polymer with 200 nm diameter fiber which when stimulated with an electric current could bend and be driven through the various curves intrinsic to the vascular system. Among many possibilities for monitoring the central nervous system this technology could eventually be used to help detect or monitor intracranial tumors [130].

Carbon nanotubes: Another author speculates about the potential role of carbon nanotubes and their role in intracranial monitoring. It could be possible that a carbon nanotubes could be embedded in a shunt that could then be used to detect specific sequences of DNA indicative of primary malignancy or recurrence [131].

Conclusion

This paper provides an overview of the current available nanobiosensor technologies being developed for diagnosis and detection of various cancers. The general themes discussed in the above sections focus on the need for reliable diagnostic tests with high sensitivity and specificity for early cancer detection. Nanotechnology-based platforms that meet these goals are still in their infancy but should become a valuable clinical and research tool in the coming years.

References

1. Hu DH, Gong P, Ma YF, Cai LT (2009) Research advancement and prospects of nanotechnology in early diagnosis and treatment of cancer. Ai Zheng 28(9):1000–1003
2. Cancer facts & figures (2012) Atlanta: American cancer society
3. Bohunicky B, Mousa S (2011) Biosensors: the new wave in cancer diagnosis nanotechnology. Sci Appl 4:1–10
4. LaRocque J, Bharali DJ, Mousa SA (2009) Cancer detection and treatment: the role of nanomedicines. Mol Biotechnol 42(3):358–366
5. Jain K (2010) A handbook of biomarkers. Springer, New York
6. Kissinger PT (2005) Biosensors-a perspective. Biosens Bioelectron 20(12):2512–2516
7. Osuwa J, Anusionwu P (2011) Some advances and prospects in nanotechnology: a review. Asian J Inf Technol 10:96–100
8. Shi J, Xiao Z, Kamaly N, Farokhzad OC (2011) Self-assembled targeted nanoparticles: evolution of technologies and bench to bedside translation. Acc Chem Res 44(10):1123–1134
9. Cuenca AG, Jiang H, Hochwald SN, Delano M, Cance WG, Grobmyer SR (2006) Emerging implications of nanotechnology on cancer diagnostics and therapeutics. Cancer 107(3):459–466
10. Seydel C (2003) Quantum dots get wet. Science 300(5616):80–81

11. Akerman ME, Chan WC, Laakkonen P, Bhatia SN, Ruoslahti E (2002) Nanocrystal targeting in vivo. Proc Natl Acad Sci U S A 99(20):12617–12621
12. Zrazhevskiy P, Sena M, Gao X (2010) Designing multifunctional quantum dots for bioimaging, detection, and drug delivery. Chem Soc Rev 39(11):4326–4354
13. Morgan NY, English S, Chen W, Chernomordik V, Russo A, Smith PD et al (2005) Real time in vivo non-invasive optical imaging using near-infrared fluorescent quantum dots. Acad Radiol 12(3):313–323
14. Gao X, Yang L, Petros JA, Marshall FF, Simons JW, Nie S (2005) In vivo molecular and cellular imaging with quantum dots. Curr Opin Biotechnol 16(1):63–72
15. Raffa V, Vittorio O, Riggio C, Cuschieri A (2010) Progress in nanotechnology for healthcare. Minim Invasive Ther Allied Technol 19(3):127–135
16. Weissleder R, Elizondo G, Wittenberg J, Rabito CA, Bengele HH, Josephson L (1990) Ultrasmall superparamagnetic iron oxide: characterization of a new class of contrast agents for MR imaging. Radiology 175(2):489–493
17. Artemov D, Mori N, Okollie B, Bhujwalla ZM (2003) MR molecular imaging of the her-2/neu receptor in breast cancer cells using targeted iron oxide nanoparticles. Magn Reson Med 49(3):403–408
18. Zhao M, Beauregard DA, Loizou L, Davletov B, Brindle KM (2001) Non-invasive detection of apoptosis using magnetic resonance imaging and a targeted contrast agent. Nat Med 7(11):1241–1244
19. Alexiou C, Jurgons R, Schmid R, Erhardt W, Parak F, Bergemann C et al (2005) Magnetic drug targeting–a new approach in locoregional tumor therapy with chemotherapeutic agents. Experimental animal studies. HNO 53(7):618–622
20. Loo C, Lin A, Hirsch L, Lee MH, Barton J, Halas N et al (2004) Nanoshell-enabled photonics-based imaging and therapy of cancer. Technol Cancer Res Treat 3(1):33–40
21. Sengupta S, Eavarone D, Capila I, Zhao G, Watson N, Kiziltepe T et al (2005) Temporal targeting of tumour cells and neovasculature with a nanoscale delivery system. Nature 436(7050):568–572
22. Loo C, Lowery A, Halas N, West J, Drezek R (2005) Immunotargeted nanoshells for integrated cancer imaging and therapy. Nano Lett 5(4):709–711
23. Liu H, Liu T, Wang H, Li L, Tan L, Fu C et al (2013) Impact of PEGylation on the biological effects and light heat conversion efficiency of gold nanoshells on silica nanorattles. Biomaterials 34(28):6967–6975
24. Hirsch LR, Stafford RJ, Bankson JA, Sershen SR, Rivera B, Price RE et al (2003) Nanoshell-mediated near-infrared thermal therapy of tumors under magnetic resonance guidance. Proc Natl Acad Sci U S A 100(23):13549–13554
25. O'Neal DP, Hirsch LR, Halas NJ, Payne JD, West JL (2004) Photo-thermal tumor ablation in mice using near infrared-absorbing nanoparticles. Cancer Lett 209(2):171–176
26. Bao G, Mitragotri S, Tong S (2013) Multifunctional nanoparticles for drug delivery and molecular imaging. Annu Rev Biomed Eng 11(15):253–282
27. Boyer D, Tamarat P, Maali A, Lounis B, Orrit M (2002) Photothermal imaging of nanometer-sized metal particles among scatterers. Science 297(5584):1160–1163
28. Jain PK, Lee KS, El-Sayed IH, El-Sayed MA (2006) Calculated absorption and scattering properties of gold nanoparticles of different size, shape, and composition: applications in biological imaging and biomedicine. J Phys Chem B 110(14):7238–7248
29. Sokolov K, Follen M, Aaron J, Pavlova I, Malpica A, Lotan R et al (2003) Real-time vital optical imaging of precancer using anti-epidermal growth factor receptor antibodies conjugated to gold nanoparticles. Cancer Res 63(9):1999–2004
30. Paciotti GF, Myer L, Weinreich D, Goia D, Pavel N, McLaughlin RE et al (2004) Colloidal gold: a novel nanoparticle vector for tumor directed drug delivery. Drug Deliv 11(3):169–183
31. Prabhakar U, Maeda H, Jain RK, Sevick-Muraca EM, Zamboni W, Farokhzad OC et al (2013) Challenges and key considerations of the enhanced permeability and retention effect for nanomedicine drug delivery in oncology. Cancer Res 73(8):2412–2417

32. Wang X, Yang L, Chen ZG, Shin DM (2008) Application of nanotechnology in cancer therapy and imaging. CA Cancer J Clin 58(2):97–110
33. Lee SK, Kim GS, Wu Y, Kim DJ, Lu Y, Kwak M et al (2012) Nanowire substrate-based laser scanning cytometry for quantitation of circulating tumor cells. Nano Lett 12(6):2697–2704
34. Zhang GJ, Ning Y (2012) Silicon nanowire biosensor and its applications in disease diagnostics: a review. Anal Chim Acta 24(749):1–15
35. Iijima S (1991) Helical microtubules of graphitic carbon. Nature 354:56–58
36. Kim SN, Rusling JF, Papadimitrakopoulos F (2007) Carbon nanotubes for electronic and electrochemical detection of biomolecules. Adv Mater 19(20):3214–3228
37. Kierny MR, Cunningham TD, Kay BK (2012) Detection of biomarkers using recombinant antibodies coupled to nanostructured platforms. Nano Rev 3. doi: 10.3402/nano.v3i0.17240. Epub 2012 Jul 23
38. Gupta AK, Nair PR, Akin D, Ladisch MR, Broyles S, Alam MA et al (2006) Anomalous resonance in a nanomechanical biosensor. Proc Natl Acad Sci U S A 103(36):13362–13367
39. Jain KK (2007) Applications of nanobiotechnology in clinical diagnostics. Clin Chem 53(11):2002–2009
40. Sanna V, Sechi M (2012) Nanoparticle therapeutics for prostate cancer treatment. Maturitas 73(1):27–32
41. Rao AR, Motiwala HG, Karim OM (2008) The discovery of prostate-specific antigen. BJU Int. 101(1):5–10
42. Charatan F (1994) FDA approves test for prostatic cancer. BMJ 309:628
43. U.S. Preventive Services (2012) Task Force Recommendation Statement. Screening for prostate cancer
44. Hill HD, Mirkin CA (2006) The bio-barcode assay for the detection of protein and nucleic acid targets using DTT-induced ligand exchange. Nat Protoc 1(1):324–336
45. Thaxton CS, Elghanian R, Thomas AD, Stoeva SI, Lee JS, Smith ND et al (2009) Nanoparticle-based bio-barcode assay redefines "undetectable" PSA and biochemical recurrence after radical prostatectomy. Proc Natl Acad Sci U S A 106(44):18437–18442
46. Jaffrezic-Renault N, Martelet C, Chevolot Y, J- Cloarec (2007) Biosensors and bio-bar code assays based on biofunctionalized magnetic microbeads. Sensors 7:589–614
47. Viator JA, Gupta S, Goldschmidt BS, Bhattacharyyal K, Kannan R, Shukla R et al (2010) Gold nanoparticle mediated detection of prostate cancer cells using photoacoustic flowmetry with optical reflectance. J Biomed Nanotechnol 6(2):187–191
48. Harisinghani MG, Barentsz J, Hahn PF, Deserno WM, Tabatabaei S, van de Kaa CH et al (2003) Noninvasive detection of clinically occult lymph-node metastases in prostate cancer. N Engl J Med 348(25):2491–2499
49. Janib SM, Moses AS, MacKay JA (2010) Imaging and drug delivery using theranostic nanoparticles. Adv Drug Deliv Rev 62(11):1052–1063
50. Kim A, Ah C, Yu H, Yang J, Baek I, Ahn C et al (2007) Ultrasensitive, label-free, and real-time immunodetection using silicon field-effect transistors. Appl Phy Lett 91(10):103901–103903
51. Zheng G, Patolsky F, Cui Y, Wang WU, Lieber CM (2005) Multiplexed electrical detection of cancer markers with nanowire sensor arrays. Nat Biotechnol 23(10):1294–1301
52. Stern E, Vacic A, Rajan NK, Criscione JM, Park J, Ilic BR et al (2010) Label-free biomarker detection from whole blood. Nat Nanotechnol 5(2):138–142
53. Malhotra R, Papadimitrakopoulos F, Rusling JF (2010) Sequential layer analysis of protein immunosensors based on single wall carbon nanotube forests. Langmuir 26(18):15050–15056
54. Chikkaveeraiah BV, Bhirde A, Malhotra R, Patel V, Gutkind JS, Rusling JF (2009) Single-wall carbon nanotube forest arrays for immunoelectrochemical measurement of four protein biomarkers for prostate cancer. Anal Chem 81(21):9129–9134

55. Lee JH, Hwang KS, Park J, Yoon KH, Yoon DS, Kim TS (2005) Immunoassay of prostate-specific antigen (PSA) using resonant frequency shift of piezoelectric nanomechanical microcantilever. Biosens Bioelectron 20(10):2157–2162
56. Hwang KS, Lee JH, Park J, Yoon DS, Park JH, Kim TS (2004) In-situ quantitative analysis of a prostate-specific antigen (PSA) using a nanomechanical PZT cantilever. Lab Chip 4(6):547–552
57. DeSantis C, Siegel R, Bandi P, Jemal A (2011) Breast cancer statistics. CA Cancer J Clin 61(6):409–418
58. Grobmyer SR, Morse DL, Fletcher B, Gutwein LG, Sharma P, Krishna V et al (2011) The promise of nanotechnology for solving clinical problems in breast cancer. J Surg Oncol 103(4):317–325
59. Colombo M, Corsi F, Foschi D, Mazzantini E, Mazzucchelli S, Morasso C et al (2010) HER2 targeting as a two-sided strategy for breast cancer diagnosis and treatment: Outlook and recent implications in nanomedical approaches. Pharmacol Res 62(2):150–165
60. Wu X, Liu H, Liu J, Haley KN, Treadway JA, Larson JP et al (2003) Immunofluorescent labeling of cancer marker Her2 and other cellular targets with semiconductor quantum dots. Nat Biotechnol 21(1):41–46
61. Chen C, Peng J, Xia HS, Yang GF, Wu QS, Chen LD et al (2009) Quantum dots-based immunofluorescence technology for the quantitative determination of HER2 expression in breast cancer. Biomaterials 30(15):2912–2918
62. Rossi LM, Shi L, Rosenzweig N, Rosenzweig Z (2006) Fluorescent silica nanospheres for digital counting bioassay of the breast cancer marker HER2/neu [correction of HER2/nue. Biosens Bioelectron 21(10):1900–1906
63. Leuschner C, Kumar CS, Hansel W, Soboyejo W, Zhou J, Hormes J (2006) LHRH-conjugated magnetic iron oxide nanoparticles for detection of breast cancer metastases. Breast Cancer Res Treat 99(2):163–176
64. Yang L, Peng XH, Wang YA, Wang X, Cao Z, Ni C et al (2009) Receptor-targeted nanoparticles for in vivo imaging of breast cancer. Clin Cancer Res 15(14):4722–4732
65. Copland JA, Eghtedari M, Popov VL, Kotov N, Mamedova N, Motamedi M et al (2004) Bioconjugated gold nanoparticles as a molecular based contrast agent: implications for imaging of deep tumors using optoacoustic tomography. Mol Imaging Biol 6(5):341–349
66. Sakamoto JH, Smith BR, Xie B, Rokhlin SI, Lee SC, Ferrari M (2005) The molecular analysis of breast cancer utilizing targeted nanoparticle based ultrasound contrast agents. Technol Cancer Res Treat. 4(6):627–636
67. Rapoport N, Gao Z, Kennedy A (2007) Multifunctional nanoparticles for combining ultrasonic tumor imaging and targeted chemotherapy. J Natl Cancer Inst 99(14):1095–1106
68. Prost AC, Menegaux F, Langlois P, Vidal JM, Koulibaly M, Jost JL et al (1998) Differential transferrin receptor density in human colorectal cancer: A potential probe for diagnosis and therapy. Int J Oncol 13(4):871–875
69. Li JL, Wang L, Liu XY, Zhang ZP, Guo HC, Liu WM et al (2009) In vitro cancer cell imaging and therapy using transferrin-conjugated gold nanoparticles. Cancer Lett 274(2):319–326
70. Lowery AR, Gobin AM, Day ES, Halas NJ, West JL (2006) Immunonanoshells for targeted photothermal ablation of tumor cells. Int J Nanomed 1(2):149–154
71. Rockall AG, Sohaib SA, Harisinghani MG, Babar SA, Singh N, Jeyarajah AR et al (2005) Diagnostic performance of nanoparticle-enhanced magnetic resonance imaging in the diagnosis of lymph node metastases in patients with endometrial and cervical cancer. J Clin Oncol 23(12):2813–2821
72. Jakub JW, Pendas S, Reintgen DS (2003) Current status of sentinel lymph node mapping and biopsy: facts and controversies. Oncologist 8(1):59–68
73. Chen SL, Iddings DM, Scheri RP, Bilchik AJ (2006) Lymphatic mapping and sentinel node analysis: current concepts and applications. CA Cancer J Clin 56(5):292–309; quiz 316-7

74. Khullar O, Frangioni JV, Grinstaff M, Colson YL (2009) Image-guided sentinel lymph node mapping and nanotechnology-based nodal treatment in lung cancer using invisible near-infrared fluorescent light. Semin Thorac Cardiovasc Surg 21(4):309–315
75. Hama Y, Koyama Y, Urano Y, Choyke PL, Kobayashi H (2007) Simultaneous two-color spectral fluorescence lymphangiography with near infrared quantum dots to map two lymphatic flows from the breast and the upper extremity. Breast Cancer Res Treat 103(1):23–28
76. Ballou B, Ernst LA, Andreko S, Harper T, Fitzpatrick JA, Waggoner AS et al (2007) Sentinel lymph node imaging using quantum dots in mouse tumor models. Bioconjug Chem 18(2):389–396
77. Takeda M, Tada H, Higuchi H, Kobayashi Y, Kobayashi M, Sakurai Y et al (2008) In vivo single molecular imaging and sentinel node navigation by nanotechnology for molecular targeting drug-delivery systems and tailor-made medicine. Breast Cancer 15(2):145–152
78. Robe A, Pic E, Lassalle HP, Bezdetnaya L, Guillemin F, Marchal F (2008) Quantum dots in axillary lymph node mapping: biodistribution study in healthy mice. BMC Cancer 8:111. doi: 10.1186/1471-2407-8-111
79. Song KH, Kim C, Cobley CM, Xia Y, Wang LV (2009) Near-infrared gold nanocages as a new class of tracers for photoacoustic sentinel lymph node mapping on a rat model. Nano Lett 9(1):183–188
80. Snyder EL, Bailey D, Shipitsin M, Polyak K, Loda M (2009) Identification of CD44v6(+)/CD24- breast carcinoma cells in primary human tumors by quantum dot-conjugated antibodies. Lab Invest 89(8):857–866
81. Talanov VS, Regino CA, Kobayashi H, Bernardo M, Choyke PL, Brechbiel MW (2006) Dendrimer-based nanoprobe for dual modality magnetic resonance and fluorescence imaging. Nano Lett 6(7):1459–1463
82. Koyama Y, Talanov VS, Bernardo M, Hama Y, Regino CA, Brechbiel MW et al (2007) A dendrimer-based nanosized contrast agent dual-labeled for magnetic resonance and optical fluorescence imaging to localize the sentinel lymph node in mice. J Magn Reson Imaging 25(4):866–871
83. Lee H, Yoon TJ, Figueiredo JL, Swirski FK, Weissleder R (2009) Rapid detection and profiling of cancer cells in fine-needle aspirates. Proc Natl Acad Sci USA. 106(30):12459–12464
84. Ali S, Coombes RC (2000) Estrogen receptor alpha in human breast cancer: Occurrence and significance. J Mammary Gland Biol Neoplasia 5(3):271–281
85. Patolsky F, Zheng G, Lieber CM (2006) Nanowire-based biosensors. Anal Chem 78(13):4260–4269
86. Shao N, Wickstrom E, Panchapakesan B (2008) Nanotube-antibody biosensor arrays for the detection of circulating breast cancer cells. Nanotechnology 19(46):465101. doi: 10.1088/0957-4484/*19*/*46*/465101. Epub 2008 Oct 21
87. Liu Z, Chen K, Davis C, Sherlock S, Cao Q, Chen X et al (2008) Drug delivery with carbon nanotubes for in vivo cancer treatment. Cancer Res 68(16):6652–6660
88. Dhar S, Liu Z, Thomale J, Dai H, Lippard SJ (2008) Targeted single-wall carbon nanotube-mediated pt(IV) prodrug delivery using folate as a homing device. J Am Chem Soc 130(34):11467–11476
89. Chen H, Han J, Li J, Meyyappan M (2004) Microelectronic DNA assay for the detection of BRCA1 gene mutations. Biomed Microdevices 6(1):55–60
90. Yang F, Jin C, Subedi S, Lee CL, Wang Q, Jiang Y et al (2012) Emerging inorganic nanomaterials for pancreatic cancer diagnosis and treatment. Cancer Treat Rev 38(6):566–579
91. Baxter NN, Whitson BA, Tuttle TM (2007) Trends in the treatment and outcome of pancreatic cancer in the United States. Ann Surg Oncol 14(4):1320–1326

92. Kumagai M, Kano MR, Morishita Y, Ota M, Imai Y, Nishiyama N et al (2009) Enhanced magnetic resonance imaging of experimental pancreatic tumor in vivo by block copolymer-coated magnetite nanoparticles with TGF-beta inhibitor. J Control Release 140(3):306–311
93. Yang L, Mao H, Cao Z, Wang YA, Peng X, Wang X et al (2009) Molecular imaging of pancreatic cancer in an animal model using targeted multifunctional nanoparticles. Gastroenterology 136(5):1514–1525.e2
94. Liong M, Lu J, Kovochich M, Xia T, Ruehm SG, Nel AE et al (2008) Multifunctional inorganic nanoparticles for imaging, targeting, and drug delivery. ACS Nano 2(5):889–896
95. Yang L, Mao H, Wang YA, Cao Z, Peng X, Wang X et al (2009) Single chain epidermal growth factor receptor antibody conjugated nanoparticles for in vivo tumor targeting and imaging. Small 5(2):235–243
96. Gu B, Xu C, Yang C, Liu S, Wang M (2011) ZnO quantum dot labeled immunosensor for carbohydrate antigen 19–9. Biosens Bioelectron 26(5):2720–2723
97. Kumar R, Roy I, Ohulchanskyy TY, Goswami LN, Bonoiu AC, Bergey EJ et al (2008) Covalently dye-linked, surface-controlled, and bioconjugated organically modified silica nanoparticles as targeted probes for optical imaging. ACS Nano 2(3):449–456
98. Vivero-Escoto JL, Taylor-Pashow KM, Huxford RC, Della Rocca J, Okoruwa C, An H et al (2011) Multifunctional mesoporous silica nanospheres with cleavable gd(III) chelates as MRI contrast agents: synthesis, characterization, target-specificity, and renal clearance. Small 7(24):3519–3528
99. Yong KT, Ding H, Roy I, Law WC, Bergey EJ, Maitra A et al (2009) Imaging pancreatic cancer using bioconjugated InP quantum dots. ACS Nano 3(3):502–510
100. Ding H, Yong KT, Law WC, Roy I, Hu R, Wu F et al (2011) Non-invasive tumor detection in small animals using novel functional pluronic nanomicelles conjugated with anti-mesothelin antibody. Nanoscale 3(4):1813–1822
101. Zaman MB, Baral TN, Jakubek ZJ, Zhang J, Wu X, Lai E et al (2011) Single-domain antibody bioconjugated near-IR quantum dots for targeted cellular imaging of pancreatic cancer. J Nanosci Nanotechnol 11(5):3757–3763
102. Law WC, Yong KT, Roy I, Ding H, Hu R, Zhao W et al (2009) Aqueous-phase synthesis of highly luminescent CdTe/ZnTe core/shell quantum dots optimized for targeted bioimaging. Small 5(11):1302–1310
103. Qian J, Yong KT, Roy I, Ohulchanskyy TY, Bergey EJ, Lee HH et al (2007) Imaging pancreatic cancer using surface-functionalized quantum dots. J Phys Chem B 111(25):6969–6972
104. Yong KT (2009) Mn-doped near-infrared quantum dots as multimodal targeted probes for pancreatic cancer imaging. Nanotechnology 20(1):015102. doi: 10.1088/0957-4484/20/1/015102. Epub 2008 Dec 5
105. Erogbogbo F, Tien CA, Chang CW, Yong KT, Law WC, Ding H et al (2011) Bioconjugation of luminescent silicon quantum dots for selective uptake by cancer cells. Bioconjug Chem. 22(6):1081–1088
106. Chang SQ, Dai YD, Kang B, Han W, Chen D (2009) Gamma-radiation synthesis of silk fibroin coated CdSe quantum dots and their biocompatibility and photostability in living cells. J Nanosci Nanotechnol 9(10):5693–5700
107. Yong KT (2010) Biophotonics and biotechnology in pancreatic cancer: cyclic RGD-peptide-conjugated type II quantum dots for in vivo imaging. Pancreatology 10(5):553–564
108. Yanez-Sedeno P, Pingarron JM (2005) Gold nanoparticle-based electrochemical biosensors. Anal Bioanal Chem 382(4):884–886
109. Khan JA, Kudgus RA, Szabolcs A, Dutta S, Wang E, Cao S et al (2011) Designing nanoconjugates to effectively target pancreatic cancer cells in vitro and in vivo. PLoS ONE 6(6):e20347
110. Eck W, Craig G, Sigdel A, Ritter G, Old LJ, Tang L et al (2008) PEGylated gold nanoparticles conjugated to monoclonal F19 antibodies as targeted labeling agents for human pancreatic carcinoma tissue. ACS Nano 2(11):2263–2272

111. Hu R, Yong KT, Roy I, Ding H, He S, Prasad PN (2009) Metallic nanostructures as localized plasmon resonance enhanced scattering probes for multiplex dark field targeted imaging of cancer cells. J Phys Chem C Nanomater Interfaces 113(7):2676–2684
112. Montet X, Weissleder R, Josephson L (2006) Imaging pancreatic cancer with a peptide-nanoparticle conjugate targeted to normal pancreas. Bioconjug Chem 17(4):905–911
113. Zhuo Y, Yuan R, Chai YQ, Hong CL (2010) Functionalized SiO2 labeled CA19-9 antibodies: a new strategy for signal amplification of antigen-antibody sensing processes. Analyst 135(8):2036–2042
114. Liu Q, Liu A, Gao F, Weng S, Zhong G, Liu J et al (2011) Coupling technique of random amplified polymorphic DNA and nanoelectrochemical sensor for mapping pancreatic cancer genetic fingerprint. Int J Nanomedicine 6:2933–2939
115. Sienel W, Dango S, Ehrhardt P, Eggeling S, Kirschbaum A, Passlick B (2006) The future in diagnosis and staging of lung cancer. Molecular techniques. Respiration 73(5):575–580
116. Barash O, Peled N, Tisch U, Bunn PA Jr, Hirsch FR, Haick H (2012) Classification of lung cancer histology by gold nanoparticle sensors. Nanomedicine 8(5):580–589
117. Peng G, Hakim M, Broza YY, Billan S, Abdah-Bortnyak R, Kuten A et al (2010) Detection of lung, breast, colorectal, and prostate cancers from exhaled breath using a single array of nanosensors. Br J Cancer 103(4):542–551
118. Gordon SM, Szidon JP, Krotoszynski BK, Gibbons RD, O'Neill HJ (1985) Volatile organic compounds in exhaled air from patients with lung cancer. Clin Chem 31(8):1278–1282
119. Ramgir N, Zajac A, Sekhar P, Lee L, Zhukov T, Bhansali S (2007) Voltammetric detection of cancer biomarkers exemplified by interleukin-10 and osteopontin with silica. J Phys Chem C 111:13981–13987
120. Peng G, Trock E, Haick H (2008) Detecting simulated patterns of lung cancer biomarkers by random network of single-walled carbon nanotubes coated with nonpolymeric organic materials. Nano Lett 8(11):3631–3635
121. Paul SP, Debono R, Walker D (2013) Clinical update: recognising brain tumours early in children. Community Pract 86(4):42–45
122. Bradbury M, Begley D, Kreuter J (2000) The blood-brain barrier and drug delivery to the CNS. Informa healthcare, Montgomery
123. Meyers JD, Doane T, Burda C, Basilion JP (2013) Nanoparticles for imaging and treating brain cancer. Nanomedicine 8(1):123–143
124. Reddy GR, Bhojani MS, McConville P, Moody J, Moffat BA, Hall DE et al (2006) Vascular targeted nanoparticles for imaging and treatment of brain tumors. Clin Cancer Res 12(22):6677–6686
125. Dilnawaz F, Singh A, Mewar S, Sharma U, Jagannathan NR, Sahoo SK (2012) The transport of non-surfactant based paclitaxel loaded magnetic nanoparticles across the blood brain barrier in a rat model. Biomaterials 33(10):2936–2951
126. Hua MY, Liu HL, Yang HW, Chen PY, Tsai RY, Huang CY et al (2011) The effectiveness of a magnetic nanoparticle-based delivery system for BCNU in the treatment of gliomas. Biomaterials 32(2):516–527
127. Kievit FM, Veiseh O, Fang C, Bhattarai N, Lee D, Ellenbogen RG et al (2010) Chlorotoxin labeled magnetic nanovectors for targeted gene delivery to glioma. ACS Nano 4(8):4587–4594
128. Bhaskar S, Tian F, Stoeger T, Kreyling W, de la Fuente JM, Grazu V et al (2010) Multifunctional nanocarriers for diagnostics, drug delivery and targeted treatment across blood-brain barrier: perspectives on tracking and neuroimaging. Part Fibre Toxicol 7:3. doi: 10.1186/1743-8977-7-3
129. Achilefu S, Dorshow RB, Bugaj JE, Rajagopalan R (2000) Novel receptor-targeted fluorescent contrast agents for in vivo tumor imaging. Invest Radiol 35(8):479–485
130. Llinás R, Walton K, Nakao M, Hunter I, Anquetil P (2005) Neuro-vascular central nervous recording/stimulating system: using nanotechnology probes. J Nanopart Res 7(2–3):111–127

131. Elder JB, Liu CY, Apuzzo ML (2008) Neurosurgery in the realm of 10(-9), part 2: applications of nanotechnology to neurosurgery–present and future. Neurosurgery 62(2):269–284. discussion 284-5
132. Liu HL, Hua MY, Yang HW, Huang CY, Chu PC, Wu JS et al (2010) Magnetic resonance monitoring of focused ultrasound/magnetic nanoparticle targeting delivery of therapeutic agents to the brain. Proc Natl Acad Sci U S A 107(34):15205–15210

Cell Compatible Arginine Containing Cationic Polymer: One-Pot Synthesis and Preliminary Biological Assessment

Nino Zavradashvili, Tamar Memanishvili, Nino Kupatadze, Lucia Baldi, Xiao Shen, David Tugushi, Christine Wandrey and Ramaz Katsarava

Abstract Synthetic cationic polymers are of interest as both nonviral vectors for intracellular gene delivery and antimicrobial agents. For both applications synthetic polymers containing guanidine groups are of special interest since such kind of organic compounds/polymers show a high transfection potential along with antibacterial activity. It is important that the delocalization of the positive charge of the cationic group in guanidine significantly decreases the toxicity compared to the ammonium functionality. One of the most convenient ways for incorporating guanidine groups is the synthesis of polymers composed of the amino acid arginine (Arg) via either application of Arg-based monomers or chemical modification of polymers with derivatives of Arg. It is also important to have biodegradable cationic polymers that will be cleared from the body after their function as transfection or antimicrobial agent is fulfilled. This chapter deals with a two-step/one-pot synthesis of a new biodegradable cationic polymer—poly(ethylene malamide) containing L-arginine methyl ester covalently attached to the macrochains in β–position of the malamide residue via the α-amino group. The goal cationic polymer was synthesized by in situ interaction of arginine methyl ester dihydrochloride with intermediary poly(ethylene epoxy succinimide) formed by polycondensation of di-*p*-nitrophenyl-*trans*-epoxy succinate with ethylenediamine. The cell compatibility study with Chinese hamster ovary (CHO) and insect Schneider 2 cells (S2) within the concentration range of 0.02–500 mg/mL revealed that the new polymer is not cytotoxic. It formed nanocomplexes with pDNA (120–180 nm in size) at low polymer/DNA weight ratios (WR = 5–10). A preliminarily transfection efficiency of the Arg-containing new cationic polymer was assessed using CHO, S2, H5, and Sf9 cells.

N. Zavradashvili (✉) · T. Memanishvili · N. Kupatadze · D. Tugushi · R. Katsarava
Institute of Chemistry and Molecular Engineering, Agricultural University of Georgia, 13 km. David Aghmashenebeli Alley 0131 Tbilisi, Georgia
e-mail: n_zavradashvili@gtu.ge

L. Baldi · X. Shen · C. Wandrey
Ecole Polytechnique Federale de Lausanne, 1015 Lausanne, Switzerland

R. Adhikari and S. Thapa (eds.), *Infectious Diseases and Nanomedicine I*, Advances in Experimental Medicine and Biology 807,
DOI: 10.1007/978-81-322-1777-0_5, © Springer India 2014

Keywords L-arginine · Polyamides · Biodegradable polycations · pDNA · Complex formation · Cytotoxicity

Introduction

Cationic polymers are widely used for various biomedical applications among which one of the most important is intracellular gene delivery—the major challenge for the success of gene therapy, which represents promising and innovative options for the treatment of serious illnesses. Safe and successful gene delivery, however, needs effective carriers that help DNA to get over biological barriers and achieve the nuclei or cytoplasm of target cells. Various cationic polymers have been broadly investigated for intracellular gene delivery (transfection agents), mainly because they easily form conjugates with negatively charged strands of DNA, condensing it into nanometer-scale structures small enough to enter cells via endocytosis [1]. Among cationic polymers those composed of the amino acid arginine (that contains the guanidine group) attracted special attention during the recent years. It was raised by the findings of Ryser and Hancock [2] that only Arg-rich, and not lysine-rich, histones were able to transport albumin into tumor cells. The biological activity of the transactivator of transcription (TAT), the peptide (**{Gly}{Arg}{Lys} {Lys} {Arg}{Arg}{Gln}{Arg}{Arg}{Arg} {Pro}{Gln}**) from HIV, is due to a high arginine content [3, 4] as well. Later, in the related studies, it was shown that the guanidine residues of arginine are essential to the ability of TAT to transfer nucleic acids, proteins, or drugs into a target cell [3–5]. Since this finding, guanidine-rich polymers such as, e.g., poly-Arg became popular transfection agents. Many Arg-based synthetic polymers including poly-Arg, however, showed rather high cytotoxicity [6–10]. Others showed good cell compatibility but low solubility in water [11] that restricts their applications.

The application as antimicrobial agents is another important scope of the application of cationic polymers. The advantages of cationic antimicrobials consist in their ability to form films and devices, in case of therapeutical applications to act in prolonged mode. Polymeric antimicrobials show much higher antimicrobial ability as compared with low-molecular-weight analogs, which is ascribed to a high local concentration of the active groups. Many effective organic antimicrobials, like transfection agents above, consist of derivatives of guanidine H_2N–(C = NH)–NH_2—a well-known and very strong organic base found in many plants and animals. A family of substituted guanidine with a wide range of properties was obtained during the last two decades [12–15]. A comprehensive chemical library of derivatives of guanidine is compiled and an antimicrobial study of these compounds was done [16]. Antimicrobial peptides (AMPs), found in all classes of living organisms, represent a large and unique group of molecules with a broad spectrum of antimicrobial, antifungal, and antiviral activity [17–19]. Unlike most conventional antibiotics, which are bacteriostatic, AMPs are rapidly bactericidal (kill bacteria). They target the microbial membrane killing rapidly via compromise membrane integrity. AMPs are composed of 12–50 amino acids, and are enriched by the cationic α-amino acid arginine [17–22]. Bactericidal polymers containing guanidine are mostly carbon-chains obtained by polymerization of the corresponding unsaturated monomers [23]. These polymers are cytotoxic and nondegradable, and less suitable for therapeutical application. As to heterochain bactericidal polymers composed of arginine, to our knowledge, only regular polypeptides $(\text{-Gly-L-Ala-L-Arg-})_n$, $(\text{-L-Lys-L-Arg-L-Ala-})_n$, $(\text{-L-Lys-L-Arg-Gly})_n$ obtained by polycondensation of activated esters of the corresponding tripeptides are reported so far [24]. The regular polypeptides showed a high bactericidal activity against *Staphylococcus aureus*, however nothing was reported about their toxicity and biodegradation.

At the same time, cytotoxicity (toward eucaryotic cells) of cationic polymers destined for the application as either gene transfection or therapeutic bactericidal agents is crucial in their practical applications. Therefore, it is relevant to discuss here briefly the cytotoxicity issue of cationic molecules or particles that are associated with cytotoxic effects [25]. For example, cationic lipids containing tertiary or quaternary amines are cytotoxic because of their interference with critical enzymes such as PKC (Protein Kinase C) [25, 26]. However, the high toxicity can be partly prevailed by delocalizing the positive charge of the cationic group. Heterocyclic rings (e.g., pyridine) or guanidine were found to be significantly less toxic when compared to the ammonium functionality, especially to the quaternary ammonium functionality [25]. Yingyongnarongkul et al. [27] synthesized guanidine-containing cationic lipids and demonstrated that these compounds were safer. It was demonstrated recently that Arg-based poly(ester amide)s, poly(ether ester amide)s, and related polymers [11, 28–31] are nontoxic within the tested concentration range (up to 2 mg/mL). At a low concentration (0.1 mg/mL) they even slightly enhanced cell growth [30].

Along with low cytotoxicity of Arg-based polymers it is desirable that they are biodegradable and cleared from the body after their function is fulfilled [32–35]. From this point of view, the Arg-based polymers with hydrolyzable ester bonds in the backbones appear promising due to the capability of these types of polymers to undergo both chemical (nonspecific) and enzyme-catalyzed hydrolysis [29–31]. However, the so far described polymers of these types were of low-molecular-weights ($\leq$3100 Da) [11, 29–31] that restrict their practical applications.

This chapter deals with the synthesis of a new biodegradable Arg-containing cationic polymer, which shows virtually no cytotoxicity and is promising for the said practical applications. The two-stage/one-pot synthetic strategy developed for this polymer and consisting in the chemical transformation of intermediary poly(ethylene epoxy-succinimide) by the in situ interaction with arginine methyl ester dihydrochloride opens a new way to synthesize a variety of Arg-enriched cationic polymers. The specific feature of the new polymers obtained according to this synthetic strategy is that they contain in the backbones neighboring amide groups (formed by 1,2-dicarboxylic acid) that support the hydrolysis of each other in acidic medium owing to the anchimeric assistance [36].

Experimental

Materials and Methods

L-Arginine, ethylenediamine, p-toluenesulfonic acid, benzene, dimethyldichlorosilane (all from Aldrich) were used as purchased. Triethylamine, TEA (from Lancaster) was purified using a standard method and distillation under atmospheric conditions. Hexamethylphosphoramide (HMPT, from Aldrich) was dried over P_2O_5 and distilled under vacuum over a fresh portion of P_2O_5. The FT-IR spectra were recorded for fine powders using a Thermo Nicolet Avatar 370 FT-IR spectrophotometer coupled with EZ OMNIC software and an Avatar Multi-Bounce Flat Plate 45° Ge. The dynamic light scattering (DLS) measurements were performed on a Nano-ZS ZEN 3600 (Malvern Instruments, UK).

Monomers

A key bis-electrophilic monomer, an activated di-*p*-nitrophenyl ester of *trans*-epoxy succinic acid N*t*ES, was synthesized by interaction of *trans*-epoxy-succinyl chloride (0.1 mol) with p-nitrophenol (0.2 mol) under the conditions of a Schotten-Baumann procedure using the two-phase system chloroform/water as reported in Ref. [37]. The product was recrystallized from acetone. The m.p. 182–184 °C was is in accordance with reported data [37].

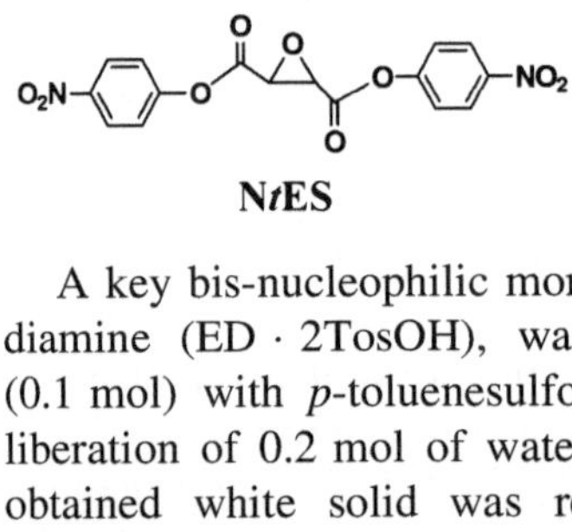

N*t*ES

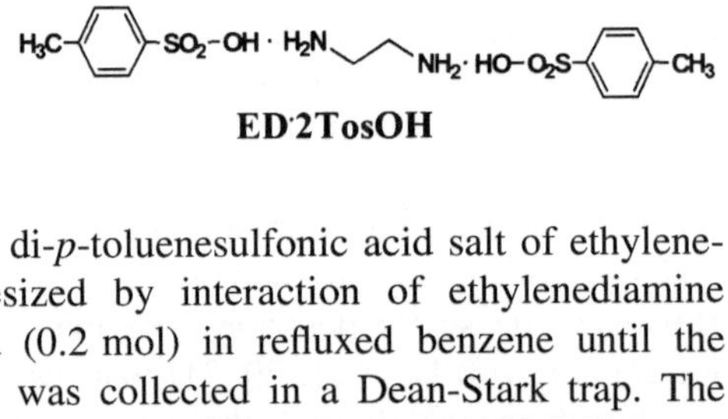

ED·2TosOH

A key bis-nucleophilic monomer, a di-*p*-toluenesulfonic acid salt of ethylenediamine (ED · 2TosOH), was synthesized by interaction of ethylenediamine (0.1 mol) with *p*-toluenesulfonic acid (0.2 mol) in refluxed benzene until the liberation of 0.2 mol of water, which was collected in a Dean-Stark trap. The obtained white solid was recrystallized from ethanol. Yield: 83.3 %, m.p. 306–308 °C.

L-Arginine methyl ester dihydrochloride (**Arg(Me) · 2HCl**) was synthesized in methanol solution in the presence of dimethyldichlorosilane.

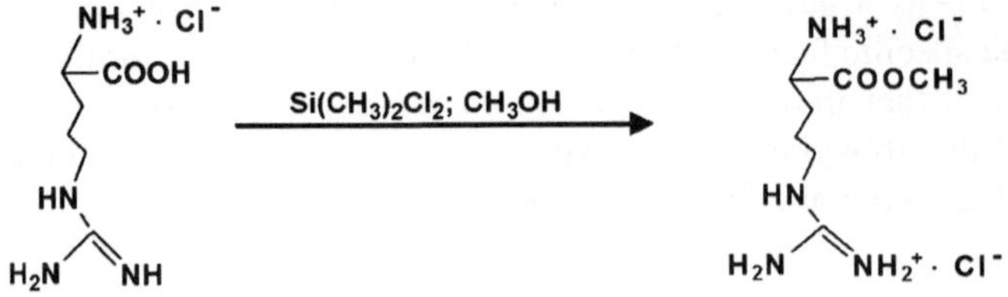

(Arg(Me)·2HCl)

Typically, to a chilled (−5 °C) mixture of Arg.HCl (0.1 mol) and 100 mL of *abs.* methanol, 10 mL of dimethyldichlorosilane were added dropwise and stirred for 2 h, then additionally for 1 h at room temperature, and finally left overnight. The excess of methanol was evaporated on a rotary evaporator. The obtained white solid was recrystallized from ethanol/diethyl ether and dried in vacuum at 50 °C. Yield: 93.6 %, the m.p. 192–193 °C coincides with reported data (190 °C, [38]). The FT-IR spectrum of Arg(Me) · 2HCl showed the ester CO absorption band at 1743 cm^{-1} and no absorption at 1666 cm^{-1}.

Polymer Synthesis and Characterization

One-Pot Synthesis

The intended Arg-containing cationic polymer was synthesized by a two-steps/one-pot synthesis consisting of in situ interaction of L-arginine methylester dichloride with the intermediate product poly(ethylene epoxy-succinimide), which was previously formed by polycondensation of the bis-electrophilic monomer N*t*ES and the bis-nucleophilic monomer ED · 2TosOH. The successive steps were the following: First, for the polycondensation, the bis-electrophilic monomer N*t*ES (2.40 g, 0.0064 mol) and the bis-nucleophilic monomer ED · 2TosOH (2.84 g, 0.0064 mol) were placed into a three-neck 50 mL round-bottom flask equipped with a stirrer. Dry HMPA (8.6 mL) and dry NEt_3 (2 mL, 0.014 mol), with a total volume of 10.6 mL corresponding to a monomer concentration 0.6 mol/L, were added and stirred gently at room temperature, for 24 h. Second, for the in situ transformation, the powdery Arg(Me) · 2HCl (2 g, 0.0076 mol) and NEt_3 (1.1 mL, 0.0076 mol) were added to the viscous homogeneous solution and stirred at 60 °C for 24 h. The reaction solution was separated from the precipitated salt (NEt_3 HCl) by filtration and poured in dry acetone. The solid product precipitated in acetone was filtered off, thoroughly washed with acetone, and dried in vacuum at 40° over P_2O_5. Yield: 89.0 %. The obtained Arg-containing cationic polymer shown in Scheme 1 and designated as *t*ES-ED-Arg(Me) was further purified to remove the unreacted Arg(Me) · 2HCl. For this, the polymer was dissolved in distilled water and dialyzed against water using Spectra/Por Biotech, CE dialysis tubing (MWCO 500) until no chloride ions were detected in the outer volume of water. The content of the dialysis bag was lyophilized yielding a yellowish-white solid.

For calibration purposes, the intermediate poly(ethylene epoxy-succinimide) designated as *t*ES-ED (Scheme 1) was separated by precipitation in water. The precipitated powdery *t*ES-ED was filtered off, thoroughly washed with acetone, and dried in vacuum at 40° over P_2O_5.

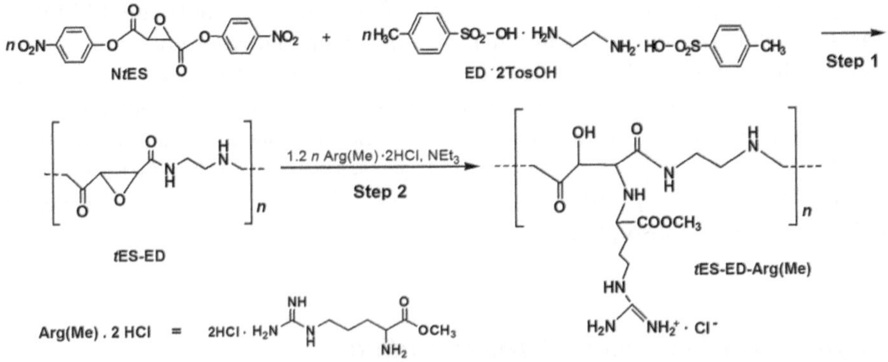

Scheme 1 The synthesis of the goal polymer tES-ED-Arg(Me)

Polymers Characterization

The structure of the goal polymer *t*ES-ED-Arg(Me) was analyzed by FT-IR. The following absorption bands were observed (cm^{-1}): 2500–3500 (wide, NH–CO + guanidine group), 1743 (CO ester), 1666 (CO amide). In the FT-IR spectrum of the intermediate *t*ES-ED, the basic absorption bands were (cm^{-1}): 3350 (NH–CO) and 1666 (CO amide), and no absorption was observed at 1743. The degree of transformation (DT) of the intermediate *t*ES-ED after the interaction with Arg(Me) · 2HCl was assessed from the ratio of the optical densities of the ester and amide absorption bands D_{1743}/D_{1666} using a calibration curve. To obtain the calibration curve, predetermined quantities of *t*ES-ED and Arg(Me) · 2HCl were mixed, thoroughly homogenized by grinding to fine powders, and a FT-IR spectrum of each mixture was recorded using Avatar Multi-Bounce Flat Plate 45° Ge. The ratio D_{1743}/D_{1666} versus the content (in mol %) of Arg(Me)·2HCl in the mixture with *t*ES-ED is plotted in Fig. 1.

Molar Mass Determination

The molar mass (MM) of the goal polymer *t*ES-ED-Arg(Me) was determined by both HPLC and DLS. The gel-chromatographic determination of the MM was done by Polymer Standards Service GmbH (Mainz, Germany) in 0.1 M NaCl/ 0.1 % (v:v) TFA using HPLC equipped with PSS SupremaMax, 10 μm, 300 Å, ID 8.0 × 300 mm columns calibrated with Dextran/Pullulane molar mass standards. DLS analysis was performed in 0.25 M acetate buffer (pH 4.5) using solutions with concentrations of 10, 15, and 20 mg/mL.

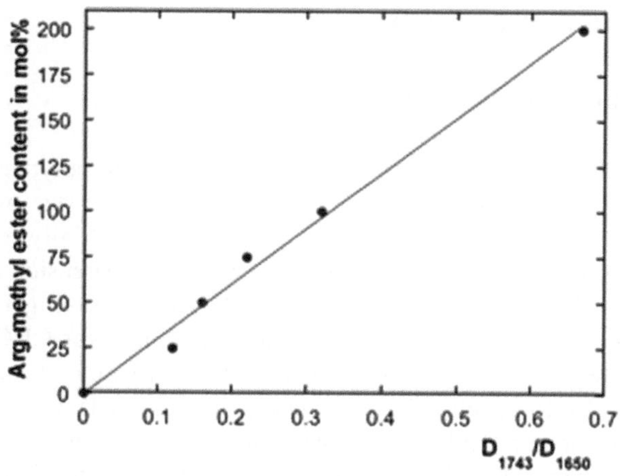

Fig. 1 Calibration *curve* Arg-methyl ester content in mol. % versus optical density ratio D_{1743}/D_{1666} ($R = 0.996$)

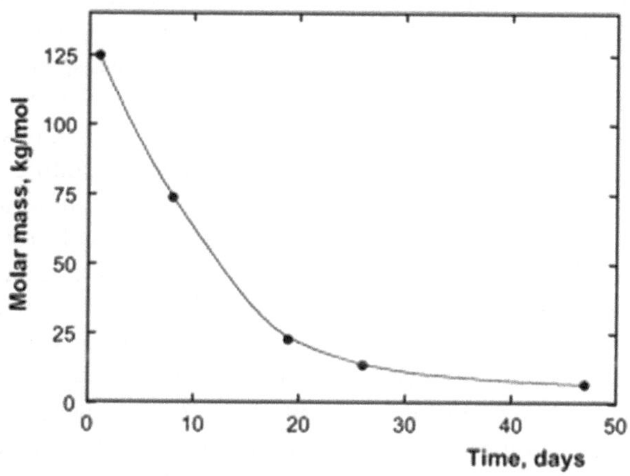

Fig. 2 The degradation *curve* of *t*ES-ED-Arg(Me) in 0.25 M acetate buffer at pH 4.5 and 37 °C

In Vitro Degradation Study

To monitor the in vitro degradation of *t*ES-ED-Arg(Me), DLS was applied. For this, the polymer *t*ES-ED-Arg(Me) was dissolved in 0.25 M acetate buffer (pH 4.5). The solution concentrations used for this study were 0.5, 1.5, and 2.5 mg/mL. The solutions were placed into the cuvettes of the Nano-ZS ZEN 3600 (Malvern Instruments, UK) and incubated at 37 °C. MM measurements were performed after predetermined periods using DLS. The obtained values were plotted versus time (Fig. 2).

Biological Study

In Vitro Cytotoxicity Study

The in vitro cytotoxicity of the new cationic polymer *t*ES-ED-Arg(Me) was evaluated in suspension cultures of Chinese hamster ovary cells (CHO) and insect Schneider 2 cells (S2), which are both industrially relevant cell lines used for the production of recombinant proteins [39]. Suspension cell cultures were maintained in TubeSpin® 50 bioreactors (TPP AG, Trasadingen, Switzerland) in serum-free culture media. ProCHO5 medium was used for CHO cells (Lonza AG, Verviers, Belgium), and Sf900 II medium (Life Technologies, Carlsbad, CA, USA) for S2 cells. Both cell cultures were kept agitated at 180 rpm in a 85 % humidified ISF1-X incubator (Kuhner AG, Birsfelden, Switzerland). CHO cell cultures were maintained at 37 °C in a 5 % CO_2 atmosphere, while S2 cells were maintained at 28 °C.

In vitro cytotoxicity was evaluated using tests that probe the metabolic activity. The cell viability was measured by Guava Viacount Reagent (Millipore, Zug, Switzerland) and performed as described in [40]. Briefly, on the day of the experiment, cells were centrifuged and resuspended in 2 mL of medium at a density of $2 - 4 \times 10^6$ cells/mL (depending on cell line) in TubeSpin® 50 bioreactor tubes. Then, the samples of new polycations were added as a solution in PBS to afford a polymer concentration in the culture medium that was varied from 0.01 to 500 mg/L. After 24 h of incubation time an aliquot of 200 μL of cell suspension was withdrawn and incubated with 180 μL Viacount Reagent in a 5 mL round-bottom polypropylene tube for 5 min. After that, the sample was diluted with 1 mL PBS and analyzed by flow cytometry in a Guava PCA-96 Easycyte™ cytofluorimeter (Guava Technologies, Hayward, USA). Cells incubated with culture medium alone were used as a control. Each experiment was carried out as three independent replicates. Viability was measured by 5,000 counts from each samples. Viability percentages after incubation with *t*ES-ED-Arg(Me) polycation for 24 h are given in Fig. 3.

Polymer/DNA Complex Formation Study

Gel Retardation Assay The DNA-binding capability of the new cation polymer *t*ES-ED-Arg(Me) was investigated by gel retardation assay with circular plasmid DNA. Polymer solutions of various concentrations were used to provide different weight ratios (WR) polymer/pDNA (pMYKEF1-EGFP-puro) [41] complexes (designated as 1, 5, 10, 50, and 100x). The polymer solutions (with $c = 0.1$, 0.5, 1.0, 5.0, and 10.0 μg/μL) and pDNA stock-solution (with $c = 0.1$ μg/μL) were filtered through a 0.45 μm sterile filter (Nylon Syringe Filters, Thermo Fisher Scientific, Pittsburgh, PA, USA). Polymer/pDNA (pMYKEF1-EGFP-puro) complexes were prepared freshly prior to use. For this 2 μL of the stock-solution of pDNA and 2 μL of each of the polymer solutions above (that provides Polymer/

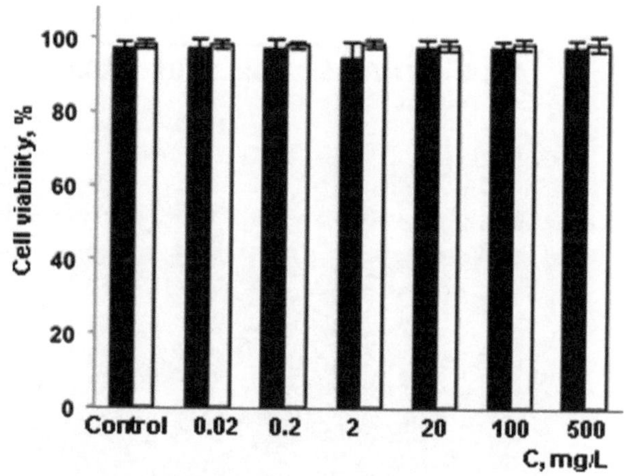

Fig. 3 The Chinese Hamster Ovary (■ CHO) and insect Schneider 2 (□ S2) cells viability after 24 h incubation with various concentration of *t*ES-ED-Arg(Me) untreated CHO and S2 cells, were used as a control

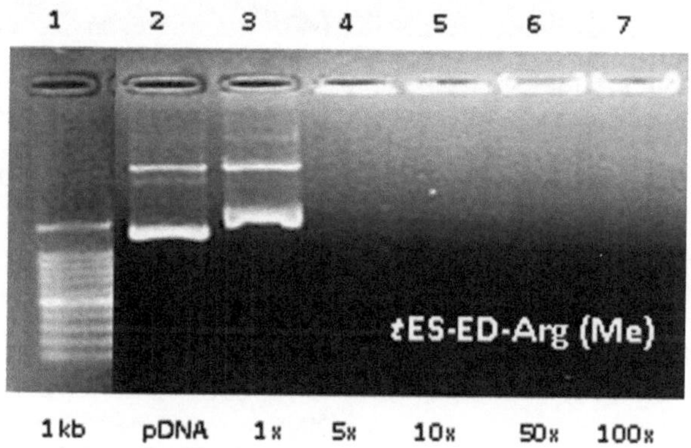

Fig. 4 Gel retardation assay of *t*ES-ED-Arg(Me) polymer *Lane 1* 1 kb DNA Ladder; *Lane 2* naked (pMYKEF1-EGFP-puro) as a control. *Lanes 3–7* different WR of polymer/pDNA complexes designated as 1, 5, 10x, etc.

pDNA WR1, 5, 10, 50, or 100x) were vortexed for 10 s and the mixtures were incubated for 30 min at room temperature. The samples were analyzed by electrophoresis on a 1 % agarose gel with Tris–acetate (TAE) running buffer at 100 V for 30–40 min. DNA bands were visualized with 0.2 μg/mL ethidium bromide staining and UV transillumination. The results are shown in Fig. 4.

DNase I digestion assay The complex formation was evaluated also by the digestion of a protective shield formed by the new cationic polymer *t*ES-ED-Arg(Me) around plasmid DNA. For the DNA digestion assay the appropriate complexes of polymer/pDNA (obtained above) were incubated for 1 h at 37 °C with 1 μL DNase I enzyme in a 20 μl reaction volume (Life Technologies). The enzymatic reaction was terminated with 4 μL of 0.5 M EDTA solution, and complexes were heated to +75 °C for 10 min. The samples were then subjected to 1 % agarose gel electrophoresis in TAE buffer at 100 V for 50–60 min and stained with 0.2 μg/mL of Gel Red (Biotium, Hayward, CA, USA). The protective shield for pDNA, was achieved even at polymer/pDNA ratio of 10x WR (Fig. 5).

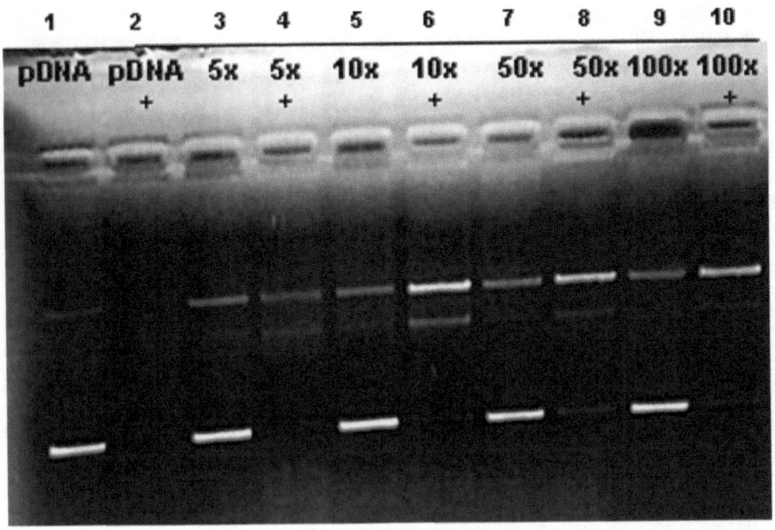

Fig. 5 Agarose (1 %) gel electrophoresis of DNase I digestion assay. *Lane 1* naked pDNA(pMYKEF1-EGFP-puro), *Lane 2* pDNA upon the addition of DNase I as a control. *Lane 3, 5, 7,* and *9* different W/R of Polymer/pDNA (designated as 5, 10x, etc.), *Lane 4, 6, 8,* and *10* the same complexes as in *Lanes 3, 5, 7,* and *9* treated with DNase I

Table 1 Particle size and zeta potential of *t*ES-ED-Arg(Me)/pDNA complexes

Weight ratio polymer/pDNA	Size (nm)	ζ (mV)
5x	181.2	24.1
10x	124.2	40.8
50x	155.7	40.0
500x	214.6	32.4

The DLS study The size and zeta potential of the obtained complexes were analyzed using a Nano-ZS ZEN 3,600 (Malvern Instruments, UK). Experiments were carried out in the standard disposable capillary cells, DTS1061 at 25 °C using the obtained above polymer/pDNA complexes. Average values were calculated from at least ten runs. The obtained data are listed in Table 1.

Preliminary Transfection Study

To evaluate the transfection efficiency of *t*ES-ED-Arg(Me) and compare it with the transfection efficiency of polyethyleneimine (PEI, Mw 25KDa), we tested other commonly used insect cell lines such as S2, H5, and Sf9, as well as CHO cells. The insect cell lines were obtained from Life Technologies and routinely cultivated as describes above for S2 (see section In Vitro Degradation Study). Transfection of CHO cells were executed as described [41]. On the day of transfection, cells were centrifuged and resuspended in 10 ml of their respective pre-warmed culture medium at the density of 4x10e + 6 cells/mL for CHO, S2, and Sf9 cells and 2x10e6 cells/ml for H5 in TubeSpin®bioreactor 50 tubes (TubeSpin; TPP, Trasadingen, Switzerland). For each culture, 300 μL of DNA (pMYKEF1-EGFP-puro for CHO and pIEx-XEGFP for insect cells) [42] and 300 μL of each of the polymer solutions above (to provide various WR polymer/pDNA) were mixed and incubated at room temperature for 30 min. The complexes were then added to the cells followed by incubation in the shaker at agitation speed of 180 rpm. The temperature and CO_2 concentration settings were as described above for CHO and insect cells, respectively. The percentage of transfected cells were measured at 2 days post-transfection using a Guava EasyCyte™ flow cytometer. Results were from 5,000 cell counts.

Discussion

For the synthesis of the new Arg-containing cationic polymer, we developed a two-steps/one-pot synthetic strategy depicted in Scheme 1. According to this new strategy, the epoxy-polyamide *t*ES-ED was synthesized by solution polycondensation of di-*p*-nitrophenyl-*trans*-epoxy succinate (N*t*ES) with di-*p*-toluenesulfonic acid salt of ethylenediamine (ED · 2TosOH) using triethylamine (NEt_3) as a *p*-toluenesulfonic acid acceptor. The epoxy-polymer formed at the first stage was interacted in situ with arginine methyl ester dihydrochloride (Arg(Me) · 2HCl) in the presence of NEt_3 resulting in the goal cationic polymer, poly(ethylene malamide) containing L-arginine methyl ester covalently attached to the macrochains in β-position of the malamide residues via α-amino groups, designated as *t*ES-ED-Arg(Me).

This synthesis relies on our recent finding that di-*p*-nitrophenyl-*trans*-epoxy succinates after the solution polycondensation with fatty diamines or their di-*p*-toluenesulfonic acid salts resulted in high-molar-mass polymers, and the obtained epoxy-polymers interacted with mono-amines under mild conditions, leading to various functionalized derivatives [37]. In the case of ethylenediamine, we used di-*p*-toluenesulfonic acid salt instead of the liquid free base (m.p. 8 °C) since the solid salt is much more convenient to purify and to manipulate. The polycondensation was carried out in HMPT in which the polycondensation proceeded homogeneously unlike in other aprotic solvents like DMF, DMA, NMP, and DMSO, in which the intermediary *t*ES-ED is insoluble, and the reaction proceeded heterogeneously. The homogeneity of the reaction is important for the subsequent stage, the in situ interaction of *t*ES-ED with Arg(Me) · 2HCl resulted in the goal cationic polymer *t*ES-ED-Arg(Me). The new polymer showed excellent water solubility due to ionogenic guanidine and *sec*-amino groups, and neutral hydroxyl groups arose after the opening epoxy-cycle (Scheme 1).

The FT-IR spectrum of the *t*ES-ED-Arg(Me) after thorough dialysis showed two absorption bands (cm^{-1}): 1743 (CO ester) and 1666 (CO amide), whereas in the FT-IR spectrum of the intermediate *t*ES-ED no absorption at 1743 was observed. This means that Arg-ester was covalently attached to the macrochains.

The FT-IR spectrum of Arg(Me) · 2HCl showed the ester CO absorption band at 1743 cm^{-1} as well. In other words, no change of the ester CO absorption band of Arg(Me) · 2HCl was observed after the attachment to the polymer chains. This allowed to use the mixtures *t*ES-ED/Arg(Me) · 2HCl for calibration purposes and quantitative determination of the degree of transformation of *t*ES-ED. The ratio D_{1743}/D_{1666} versus the content of Arg(Me) · 2HCl (in mol %) in the mixture is given in Fig. 1. The degree of transformation of *t*ES-ED into *t*ES-ED-Arg(Me) determined by this calibration curve was about 75 %.

The HPLC molar mass determination revealed that the obtained *t*ES-ED-Arg(Me) has a rather high molar mass of $M_w = 19.4$ kg/mol. The MM of this polymer was also determined by DLS within the concentration range of 10–20 mg/mL and was found as 21.6 kg/mol, which coincides well with the HPLC measurement.

The in vitro biodegradation of *t*ES-ED-Arg(Me) was performed at 37 °C in 0.25 M acetate buffer with pH 4.5 within the concentration range of 0.5–2.5 mg/mL and was monitored by DLS. At lower concentrations, the apparent MM determined by DLS was too high (130 kg/mol) due to polyelectrolyte effects. However, this range of concentration was found to be useful to monitor the biodegradation. According to the study performed, the cationic polymer *t*ES-ED-Arg(Me) was subjected to degradation under the given conditions, as it was expected taking into account that the polymer contains vicinal amide groups, which support the hydrolysis of each other in acidic medium owing to the anchimeric assistance [36].

The formation of nanocomplexes between the new cationic polymer *t*ES-ED-Arg(Me) and pDNA was studied using both gel retardation and DNase I digestion assays, and DLS. The results of the gel retardation assay given in Fig. 4 shows that the tested *t*ES-ED-Arg(Me) was able to bind pDNA and thus retarded DNA mobility during electrophoresis. A high binding potential of *t*ES-ED-Arg(Me) was expected due to Arg-residues, the concentration of which in the polymeric chains is high enough owing to a short elemental link. This assumption was confirmed by the formation of the complexes at rather low polymer/pDNA weight ratio (5x).

The complex formation was evaluated also by the digestion of a protective shield formed by the new cationic polymer *t*ES-ED-Arg(Me) around plasmid DNA. The date given in Fig. 5 shows the protective shield for pDNA, was achieved even at polymer/pDNA ratio of 10x WR.

The DLS study shows that the complexes formed after binding the polymer with pDNA are in the nanometer range (124–215 nm) (Table 1) and their zeta potential (ζ) is within of 24–41 mV that means the complexes are stable. The smaller complexes were formed at polymer/pDNA WR 10x.

To evaluate the transfection efficiency of *t*ES-ED-Arg(Me), cellular uptake of polymer/pDNA (pMYKEF1-EGFP-puro and pIEx-XEGFP) complexes at various WR was studied. For a comparative study, the well-known transfection agent poly(ethyleneimine) (PEI, $M_W = 25$ kg/mol) was used. The transfection efficiency was examined with several cell lines including Mammalian cells CHO and Insect cells S2, H5, and Sf9. The study was done under a protocol optimal for PEI. Under

Table 2 Transfection efficiency of *t*ES-ED-Arg(Me) versus PEI with various cell lines

Cells	*t*ES-ED-Arg(Me)		PEI	
	Polymer/pDNA WR	Transfected cells (%)	Polymer/pDNA WR	Transfected cells (%)
CHO	12.5x	1.7	5x	63.0
S2	10.0x	2.0	2x	51.3
H5	100.0x	4.6	3x	54.3
Sf9	60.0x	5.6	1.5x	16.6

these conditions, transfection efficiency study with *t*ES-ED-Arg(Me) showed the results given in Table 2. It is visible that the transfection efficiency of the new cationic polymer is not as high as the efficiency of PEI. That could be ascribed to the transcription protocol that might be not optimal for *t*ES-ED-Arg(Me). The optimization of the transfection protocol as well as further modification of the new Arg-containing polymer is in progress. We plan also to study the antibacterial activity of *t*ES-ED-Arg(Me) and related cationic polymers obtained via the one-pot synthetic strategy developed.

Conclusions

The developed new two-stage/one-pot synthetic strategy yielded the new Arg-containing cationic polymer *t*ES-ED-Arg(Me) of relatively high molar mass, which showed high water solubility, lack of cytotoxicity and formed nanocomplexes with pDNA at low weight ratios, 5–10x. These results were confirmed by gel retardation and DNase I digestion assays, as well as dynamic light scattering studies. The complexes were rather small in size (124–214 nm) and had positive zeta potentials (24–41 mV). The added value of the cationic polymer obtained via the new scheme is its biodegradability, since biodegradable transfection agents that will be cleared from the body after their function is fulfilled, are of special interest. So far, the new polymer showed weak transfection efficiency. It is suggested that this can be improved under optimal conditions. It is anticipated that the polymer will be a much more perfect and effective carrier for intracellular gene delivery compared to the existing cationic polymers such as PEI, poly(L-lysine), poly(arginine). Better cationic polymers are also promising, for example, as bactericidal agents, surfactants, or additives to biodegradable scaffolds for cell cloning.

Acknowledgments This study was supported by the Swiss National Science Foundation (SNF), grants SCOPES IZ73Z0_128071 and SCOPES IZ76ZO_147554. We thank Mrs. P. Toidze for DLS measurements.

References

1. Luo D, Saltzman WM (2000) Synthetic DNA delivery systems. Nat Biotechnol 18:33–37
2. Ryser H, Hancock R (1965) Histones and basic polyamino acids stimulate the uptake of albumin by tumor cells in culture. Science 150:501–503
3. Mitchell DJ, Kim DT, Steinman L, Fathman CG, Rothbard JB (2000) Polyarginine enters cells more efficiently than other polycationic homopolymers. J Pept Res 56(5):318–325
4. Calnan BJ, Tidor B, Biancalana S, Hudson D, Frankel AD (1991) Arginine-mediated RNA recognition: the arginine fork. Science 252:1167–1171
5. Rothbard JB, Garlington S, Lin Q, Kirschberg T, Kreider E, Wender PA, Khavari PA (2000) Conjugation of arginine oligomers to cyclosporin a facilitates topical delivery and inhibition of inflammation. Nat Med 6:1253–1257
6. Holowka EP, Sun VZ, Kamei DT, Deming TJ (2007) Polyarginine segments in block copolypeptides drive both vesicular assembly and intracellular delivery. Nat Mater 6:52–57
7. Rothbard JB, Jessop TC, Wender PA (2005) Adaptive translocation: the role of hydrogen bonding and membrane potential in the uptake of guanidinium-rich transporters into cells. Adv Drug Deliv Rev 57:495–504
8. Futaki S (2005) Membrane-permeable arginine-rich peptides and the translocation mechanisms. Adv Drug Deliv Rev 57:547–558
9. Brooks H, Lebleu B, Vives E (2005) Tat peptide-mediated cellular delivery: back to basics. Adv Drug Deliv Rev 57:559–577
10. Torchilin VP (2006) Recent approaches to intracellular delivery of drugs and DNA and organelle targeting. Annu Rev Biomed Eng 8:343–375
11. Yamanouchi D, Wu J, Lazar AN, Kent KC, Chu CC, Liu B (2008) Biodegradable arginine-based poly(ester-amide)s as non-viral gene delivery reagents. Biomaterials 29(22): 3269–3277
12. Weber E, Keana J (1993) N, N'—disubstituted guanidines and their use as excitatory amino acid antagonists. US. Patents 3,252,816 (1990); 5,190,976 (1993)
13. Graham JD, Lain-Yen H, Sharad M (1999) Therapeutic substituted guanidines. US Patent 5,922,772
14. Goldin SM (1997) Substituted adamantyl guanidines and methods of use there of US patent No.5,637623
15. Manetti F, Castagnolo D, Raffi F, Zizzari AT, Rajamäki S, D'Arezzo S, Visca P, Cona A, Fracasso ME, Doria D, Posteraro B, Sangui netti M, Fadda G, Botta M (2009) Synthesis of new linear guanidines and macrocyclic amidinourea derivatives endowed with high antifungal activity against Candida spp. and Aspergillus spp. J Chem Med 52(23):7376–7379
16. Hensler ME, Bernstein G, Nizet V, Nefzi A (2006) Pyrrolidine bis-cyclic guanidines with antimicrobial activity against drug-resistant gram-positive pathogens identified from a mixture-based combinatorial library. Bioorg Med Chem Lett 16(19):5073–5079
17. Yeaman MR, Yount NY (2003) Mechanisms of antimicrobial peptide action and resistance. Pharmacol Rev 55(1):27–55
18. Zasloff M (1992) Antibiotic peptides as mediators of innate immunity. Curr Opin Immunol 4:3–7
19. Zasloff M (2002) Antimicrobial peptides of multicellular organisms. Nature (London) 415:389–395
20. Sitaram N, Nagaraj R (2002) Host-defense antimicrobial peptides: importance of structure for activity. Curr Pharm Des 8(9):727–742
21. Hancock RE, Chapple DS (1999) Peptide antibiotics. Antimicrob Agents Chemother 43:1317–1323
22. Hancock RE, Lehrer R (1998) Cationic peptides: a new source of antibiotics. Trends Biotechnol 16:82–88
23. Sivov NS (2006) Biocide guanidine containing polymers: synthesis, structure and properties. CRC Press, Taylor & Francis Group, Boca raton

24. Kasymova GF, Burichenko VK, Meitus ÉE, Poroshin KT (1973) Lysine- and arginine-containing polypeptides of regular structure and their bactericidal properties. Chem Nat Compd 9(1):79–81
25. Lu H, Zhang S, Wang B, Cui S, Yan JJ (2006) Inhibition of protein kinase C by cationic amphiphiles. J Controll Rel 114(1):100–109
26. Bottega R, Epand RM (1992) Inhibition of protein kinase C by cationic amphiphiles. Biochemistry 31(37):9025–9030
27. Yingyongnarongkul B, Howarth M, Elliott T, Bradley M (2004) Solid-phase synthesis of 89 polyamine-based cationic lipids for DNA delivery to mammalian cells. Chem Eur J 10(2):463–473
28. Katsarava R, Tugushi D, Torchilin VP, Memanishvili T, Gverdtsiteli M, Kupatadze N (2010) Biodegradable L-Arginine based cationic poly(Ether-Ester-Amide)s, poly(Ether-Ester-Urethanes) and poly(Ether-Ester-Ureas) and methods of synthesis thereof. Georgian patent application, AP2009011558. (Positive decision)
29. Memanishvili T, Kupatadze N, Tugushi D, Torchilin VP, Katsarava R (2010) Biodegradable arginine-based polymers with PEG-like backbones as potential non-viral gene delivery system. In: 1st Russian-hellenic symposium with international participation and Yong's scientist school "Biomaterials and bionanomaterials: recent advances and safety-toxicology issues", Iraklion, 3–9 May 2010
30. Memanishvili T, Bedinashvili M, Tugushi D, Torchilin VP, Katsarava R (2010) The Study on the cytotoxicity of new biodegradable Arginine-based poly(Ether-Esther-Urethane cation)s on 4T1 cells. Georgian Eng News 4(56):93–98
31. Memanishvili T, Zavradashvili N, Kupatadze N, Tugushi D, Gverdtsiteli M, Torchilin VP, Wandrey C, Baldi L, Manoli SS, Katsarava R. Arginine-based biodegradable ether-ester polymers of low cytotoxicity as potential gene carriers (submitted to Biomacromolecules).
32. Vert M (2009) Degradable and bioresorbable polymers in surgery and in pharmacology: beliefs and facts. J Mater Sci Mater Med 20:437–446
33. Lynn DM, Langer R (2000) Degradable poly(β-amino esters): synthesis, characterization, and self-assembly with plasmid DNA. J Am Chem Soc 122:10761–10768
34. Akinc A, Lynn DM, Anderson DG, Langer R (2003) Parallel synthesis and biophysical characterization of a degradable polymer library for gene delivery. J Am Chem Soc 125:5316–5323
35. Forrest M, Koerber J, Pack D (2003) A degradable polyethylenimine derivative with low toxicity for highly efficient gene delivery. Bioconjug Chem 14:934–940
36. Cohen T, Lipowitz J (1964) Acid catalyzed amide hydrolysis assisted by neighboring amide group. J Amer Chem Soc 86:5611–5616
37. Zavradashvili N, Jokhadze G, Gverdtsiteli M, Otinashvili G, Kupatadze N, Gomurashvili Z, Tugushi D, Katsarava R (2013) Amino acid based epoxy-poly(Ester Amide)s—a new class of functional biodegradable polymers: synthesis and chemical transformations. J Macromol Sci Part A Pure Appl Chem 50(5):449–465
38. ChemicalBook. http://www.chemicalbook.com/
39. Wurm FM (2004) Production of recombinant protein therapeutics in cultivated mammalian cells. Nat Biotechnol 22(11):1393–1398
40. Kadlecova Z, Baldi L, Hacker D, Wurm FM, Klok HA (2012) Comparative study on the in vitro cytotoxicity of linear, dendritic, and hyperbranched polylysine analogues. Biomacromolecules 13(10):3127–3137
41. Rajendra Y, Kiseljak D, Baldi L, Hacker DL, Wurm FM (2011) A simple high-yielding process for transient gene expression in CHO cells. J Biotechnol 153:22–26
42. Shen X, Michel PO, Xie Q, Hacker DL (2011) Wurm FW (2011) Transient transfection of insect Sf-9 cells in TubeSpin® bioreactor 50 tubes. BMC Proc 5(suppl 8):P37

Neuroinvasive and Neurotropic Human Respiratory Coronaviruses: Potential Neurovirulent Agents in Humans

Marc Desforges, Alain Le Coupanec, Élodie Brison, Mathieu Meessen-Pinard and Pierre J. Talbot

Abstract In humans, viral infections of the respiratory tract are a leading cause of morbidity and mortality worldwide. Several recognized respiratory viral agents have a neuroinvasive capacity since they can spread from the respiratory tract to the central nervous system (CNS). Once there, infection of CNS cells (neurotropism) could lead to human health problems, such as encephalitis and long-term neurological diseases. Among the various respiratory viruses, coronaviruses are important pathogens of humans and animals. Human Coronaviruses (HCoV) usually infect the upper respiratory tract, where they are mainly associated with common colds. However, in more vulnerable populations, such as newborns, infants, the elderly, and immune-compromised individuals, they can also affect the lower respiratory tract, leading to pneumonia, exacerbations of asthma, respiratory distress syndrome, or even severe acute respiratory syndrome (SARS). The respiratory involvement of HCoV has been clearly established since the 1960s. In addition, for almost three decades now, the scientific literature has also demonstrated that HCoV are neuroinvasive and neurotropic and could induce an overactivation of the immune system, in part by participating in the activation of autoreactive immune cells that could be associated with autoimmunity in susceptible individuals. Furthermore, it was shown that in the murine CNS, neurons are the main target of infection, which causes these essential cells to undergo degeneration and eventually die by some form of programmed cell death after virus infection. Moreover, it appears that the viral surface glycoprotein (S) represents an important factor in the neurodegenerative process. Given all these properties, it has been suggested that these recognized human respiratory

M. Desforges (✉) · A. Le Coupanec · É. Brison · M. Meessen-Pinard · P. J. Talbot
Laboratory of Neuroimmunovirology, INRS-Institut Armand-Frappier, Institut national de la recherche scientifique, Université du Québec, 531 boulevard des Prairies, Laval, QC H7V IB7, Canada
e-mail: marc.desforges@iaf.inrs.ca

P. J. Talbot
e-mail: pierre.talbot@iaf.inrs.ca

R. Adhikari and S. Thapa (eds.), *Infectious Diseases and Nanomedicine I*,
Advances in Experimental Medicine and Biology 807,
DOI: 10.1007/978-81-322-1777-0_6, © Springer India 2014

pathogens could be associated with the triggering or the exacerbation of neurological diseases for which the etiology remains unknown or poorly understood.

Keywords Respiratory viral infection · Coronavirus · Neuroinvasion · CNS infection · Neurological diseases

Introduction

Viral infections of the respiratory tract represent a major problem for human and animal health around the world. These respiratory infections induce the most common illnesses [1] and are a leading cause of morbidity and mortality in humans worldwide, especially children, the elderly, and immune-compromised individuals [2–4]. On a statistical basis, children represent a highly susceptible group, as they may experience multiple infections each year until they reach the age of 10 [5]. Respiratory viral infections also represent an economic threat for agriculture, especially in cattle, swine, and poultry production. The idea that viruses can cause respiratory tract infections has been demonstrated since the early 1930s [2]. Nevertheless, with the help of modern diagnostic tools, a significant number of new respiratory viruses have been discovered since the beginning of the twenty-first century and it is estimated that there are about 200 antigenically distinct viruses able to cause infection of the respiratory tract, especially in infants and children [6]. In fact, it is now believed that viruses cause 95 % of respiratory diseases in children and infants, and about 30–40 % in the elderly [2].

Although the airway epithelial cells in the respiratory tract represent a first line of defense against pathogens, they can be targeted by several different respiratory viruses that can infect them as a way to penetrate the human host. Several infections are self-limited and the infection remains local as the virus is cleared by the immune system in the respiratory tract with minimal clinical consequences. However, in some circumstances, viruses can avoid the immune response and cause more severe respiratory diseases [1] or even spread to other tissues, including the central nervous system (CNS), where they could induce other types of pathologies [7].

Neuroinvasive and Neurotropic Viruses: Associated Neuropathologies

Over the years, the CNS has been shown to represent a frequent site of viral infection. Using different routes of entry, several different viruses, including respiratory pathogens, have been shown to be able to penetrate the CNS (neuroinvasion), where they can infect neurons and glial cells (neurotropism) and possibly induce or participate in the induction of neurological diseases (neurovirulence) [8]. In humans, a long list of viruses possess these "neuroproperties" and infection often leads to acute encephalitis, which can be fatal depending on virus tropism [9]. Rabies virus [10], herpes simplex virus (HSV) [11], and arthropod-borne flaviviruses [12] can induce encephalitis in humans. Chronic human neurological diseases may also be linked to viral infection. In acquired immunodeficiency syndrome (AIDS) dementia and related disorders, human immunodeficiency virus (HIV) induces neurodegeneration [13], which can result in motor dysfunctions and possibly cognitive impairments [14]. Progressive multifocal leukoencephalopathy (PML) is a human demyelinating disease [15] where prolonged immunosuppression leads to reactivation of latent polyoma JC virus (JCV) [16]. Subacute sclerosing panencephalitis (SSPE), a progressive fatal neurological disease, is caused by CNS persistence of measles virus [17]. Human T-cell lymphotrophic virus (HTLV-1) causes progressive tropical spastic paraperesis/HTLV-1-associated myelopathy (PTSP/HAM) in 1–2 % of infected individuals [18] and HSV-1 and human herpes virus 6 (HHV-6) were proposed to cause or exacerbate Alzheimer's disease (AD) [19].

As mentioned above, respiratory viral agents also have the capacity to invade the CNS where they will infect resident cells and potentially be neurovirulent in inducing a neuropathology. Several of these recognized respiratory pathogens can gain access to the CNS, where they can eventually cause health problems in humans.

Respiratory Viruses with Neuroinvasive and Neurotropic Properties: Associated Neuropathologies

Respiratory syncytial virus (RSV), the most common pathogen to cause lower respiratory tract infection in infants worldwide [20], is one such neuroinvasive respiratory agent that has been detected in the cerebrospinal fluid (CSF) of patients [21, 22] and that was associated with convulsions [23], febrile seizures, and encephalitis [24]. Furthermore, it was recently shown that RSV can spread from the airways to the CNS in mice after intranasal inoculation, and that it induces behavioral and cognitive impairments [25].

Measles virus (MV) from the *Paramyxoviridae* family is another common virus that causes a disease of the respiratory airways associated with fever, cough, and congestion. However, MV infection also induces other symptoms including a characteristic rash and Koplik's spots [26] in the oral mucosa. One of two most important sequelae associated with MV infection is immunosuppression, which facilitates infection by opportunistic bacteria or parasites that can lead to pneumonia or diarrhea. A second type of rare but significant sequelae is long-term CNS disease [26]. Postinfectious encephalomyelitis (PIE) or acute disseminated encephalomyelitis (ADEM) occurs in 1 of 1,000 measles cases in children and adolescents. Measles inclusion body encephalitis (MIBE) is a second CNS complication that can arise after a MV infection in immune-compromised patients. Finally, subacute sclerosing panencephalitis (SSPE) is a third form of CNS disease associated with MV infection. It is a slow progressive neurological disease that appears 6–10 years after infection in about 4–11 cases per 100,000 cases of measles (for review see [27]).

Hendra virus (HeV) and Nipah virus (NiV) are two members of the genus Henipavirus (HNV), also from the *Paramyxoviridae* family. They represent important emerging viruses that were discovered in the 1990s [28]. Fruit bats are the natural reservoir and both viruses can be transmitted to humans from intermediate reservoirs such as pigs and horses [29]. The HNV causes acute and severe respiratory disease in humans, including necrotizing alveolitis with hemorrhage, pulmonary edema, and pneumonia [28]. Neurological signs of pathology include confusion, motor deficits, seizures, febrile encephalitic syndrome, and reduced level of consciousness. Moreover, neuropsychiatric sequelae have been reported but it is not known whether postinfectious encephalomyelitis occurs following infection [29]. The use of animal models showed that the main route of entry into the CNS is the olfactory nerve [30].

Influenza virus comes in three types: A, B, and C. Types A and B cause the flu syndrome, characterized by chills, fever, headache, sore throat, and muscle pains [31], and are responsible for seasonal epidemics that affect 3–5 million humans, of which 250,000–500,000 cases are lethal each year [32]. Although influenza type C virus is less frequent in humans, it is also distributed around the world where it mainly infects infants and young children; most of the time infection results in mild upper respiratory tract illnesses. However, complications associated with lower respiratory tract infection do occur in children under 2 years of age [33, 34]. Human influenza A virus is of particular interest because of its capacity to recombine and rearrange with its avian and porcine counterparts to generate new emerging viruses that are introduced into the human population and that may lead to pandemic outbreaks associated with significant morbidity and mortality, as it was observed at least four times during the twentieth century and already once in the twenty-first century [32, 35]. Influenza virus type A is highly contagious and, even though most infections are localized to the upper respiratory tract, some more severe cases may result in pneumonia [36] and even complications involving the CNS [37]. Several studies have shown that influenza A can be associated with encephalitis, Reye's syndrome, febrile seizure, acute necrotizing encephalopathy, and possibly acute disseminated encephalomyelitis (ADEM) in humans [31, 38–41]. Making use of murine models, it has also been shown that influenza A virus could reach the CNS through the olfactory nerve route and alter hippocampal morphology or expression of synaptic regulatory genes while impairing cognition and emotional behavior [42, 43]. Influenza A virus was also described as a factor which may increase the risk of Parkinson's disease (PD) [37].

Among the different respiratory viruses, coronaviruses are important pathogens of humans and animals, causing a range of symptoms, including in the CNS.

Coronaviruses

Coronaviruses, a family of enveloped positive-stranded RNA viruses with a characteristic crown-shaped appearance, are widespread in nature and can infect several different species [44], in which they cause mainly respiratory and enteric pathologies, with neurotropic and neuroinvasive properties in various hosts including humans, cats, pigs, rodents, and fowl [45–48].

They are taxonomically grouped in the family *Coronaviridae*, within the order *Nidovirales,* and they are classified within four different genera, namely Alpha, Beta Gamma, and Deltacoronaviruses [49, 50]. They form a group of enveloped viruses that have the largest genome among RNA viruses. This non-segmented 30 kb positive single-stranded polyadenylated RNA possesses 4 or 5 genes encoding structural proteins (S, E, M, N; HE for the genus Betacoronaviruses) and several genes encoding non-structural proteins.

The spike protein (S) is a type-1 glycosylated transmembrane protein responsible for the recognition of the cellular receptor used by the virus to infect a susceptible cell. The envelope (E) protein is a small structural protein anchored in the viral envelope, and which has a role in the assembly of the virion; it appears to be responsible for the adequate curving of the viral envelope. The membrane (M) protein possesses three transmembrane domains and interacts with all the other structural viral proteins and therefore helps to shape and maintain the virion structure. The nucleocapsid (N) protein associates with the viral genome and plays an essential role in encapsidating it into a helical nucleocapsid within the viral particle. The hemagglutinin-esterase (HE) is only present in most species of the betacoronavirus genus. Like the S protein, it is a transmembrane protein which forms homodimers and which interacts with different types of sialic acid, associated with an apparent role in hemagglutination. It also possesses an acetyl-esterase function, which may be important early during infection or during the release of viral particles from the infected cells at the end of the replication cycle of the betacoronaviruses [48].

Animal Coronaviruses: Pathogens of the Respiratory Tract and the Central Nervous System

As mentioned above, coronaviruses are widespread in nature and can infect several different animal species, in which very often they are both respiratory and enteric pathogens. Although several animal coronaviruses induce severe enteric diseases, they also often reach the respiratory tract, where they can be associated with mild to severe diseases. Some examples of coronaviruses that can infect livestock or poultry, in which they have a respiratory tropism, are transmissible gastrointestinal virus (TGEV) and its associated S protein deletion mutant: porcine respiratory coronavirus (PRCoV) and porcine hemagglutinating encephalitis virus (PHEV), which infect swine; bovine coronavirus (BCoV), which infects cattle; and infectious bronchitis virus (IBV), which infects chicken [51]. Several of these animal coronaviruses cause severe economic burden to poultry, swine, and cattle industries worldwide. Other animal coronaviruses that have a respiratory tropism are feline coronavirus (FCoV); canine respiratory coronavirus (CRCoV); and bat CoV, which infects the bat species *Miniopterus* [51].

Among the respiratory animal coronaviruses, FCoV and PHEV have been associated with neurological diseases. Neurological symptoms may occur in cats infected with a highly virulent FCoV variant, designated FIPV [52, 53]. Neurological disease appears partially immune-mediated and may result in uncontrolled secretion of cytokines [54] that leads to diverse pathological manifestations including meningitis [55] and even spinal cord involvement [56]. Furthermore, there is often a small amount of infectious FCoV present in brain tissue [52]. The PHEV was isolated from the brains of suckling pigs suffering from encephalomyelitis several years ago in Canada [57] and the disease could be reproduced experimentally in piglets following intranasal inoculation [58]. After oronasal infection, it was shown that the virus first infects epithelial cells of the respiratory tract and small intestine and, using retrograde neuronal spreading via peripheral nerves, was able to enter the CNS [59]. More recently, the neuroinvasiveness and neurotropism of the virus were again demonstrated in a murine model, where PHEV induced a poor inflammatory reaction in the CNS and infected cells showed no cytopathological changes [60].

The last, but not the least, example of another animal coronavirus, i.e., neuroinvasive, neurotropic, and neurovirulent, is mouse hepatitis virus (MHV). This virus is a subspecies of the species murine coronavirus (MuCoV) [49], which represents a collection of viral strains with different tropism, including respiratory for the MHV-1 strain, and neurotropic for the MHV-JHM and MHV-A59 strains. In susceptible mice, these two latter strains of MHV induce a demyelinating disease that resemble human multiple sclerosis (MS) [61].

Coronaviruses are all molecularly related in structure and mode of replication [62]. Therefore, the close structural and biological relatedness of HCoV to the neurotropic animal coronaviruses has led to speculation about possible involvement of HCoV in neurological diseases. Till now, no clear specific association has ever been made between coronaviruses and any known human neuropathology. However, HCoV-229E and HCoV-OC43 [63–66], as well as SARS-CoV [67, 68], were shown to be neuroinvasive and neurotropic.

Human Coronaviruses: Respiratory Pathogens

Human coronaviruses (HCoV) usually infect the upper respiratory tract, where they are mainly associated with common colds. However, in more vulnerable populations such as newborns, infants, the elderly, and immune-compromised individuals, they can also reach the lower respiratory tract, where they could instead be associated with pneumonia, exacerbations of asthma, respiratory distress syndrome or even severe acute respiratory syndrome (SARS) [44, 69].

Ever since their discovery in the late 1960s, coronaviruses able to infect humans were neglected by the international medical community. However, when a variant emerged from animals in Southeast Asia to cause the first pandemic of the twenty-first century: the severe acute respiratory syndrome or SARS, these apparently innocuous viruses suddenly became "more interesting." Indeed, the 2002–2003 SARS pandemic was caused by a coronavirus variant that appears to have emerged from a bat reservoir to infect palm civets, sold live in open markets, which served as intermediate reservoirs before crossing into humans. Moreover in the fall of 2012, 10 years after the SARS episode, the World Health Organization (WHO) warned the international medical community, that a SARS-like disease affected individuals that traveled from the Arabian Peninsula to the United Kingdom, which may indicate a possible resurgence of SARS. However, using molecular sequencing, it was rapidly shown that this new respiratory coronavirus was genetically different from SARS-CoV, underlining the importance of a molecular approach in making a viral diagnostic. It is now recognized that the new epidemic is caused by a new coronavirus from the genus Betacoronavirus that was first named HCoV-EMC (for Human Coronavirus—Erasmus Medical Center), human betacoronavirus 2c, and NCoV or nCoV (for novel Coronavirus), and that is now known under the official name MERS-CoV: the Middle-East Respiratory Syndrome Coronavirus [50]. As of August 30, 2013, WHO indicated that the MERS-CoV has spread to eight different countries, where 108 laboratory-confirmed cases of individuals have been identified as infected by the MERS-CoV, with 50 deaths [70].

Although coronaviruses that infect humans are mainly recognized respiratory pathogens, as they usually first target respiratory and mucosal surfaces, infectious particles, antigens or RNA, were detected in other tissues than the respiratory tract, including the CNS.

Human Coronaviruses in the CNS

The detection of HCoV RNA in human brain samples clearly demonstrates that these respiratory pathogens are naturally neuroinvasive in humans and suggests that they establish a persistent infection in human CNS [63]. Furthermore, we have shown that these viruses are able to establish a persistent infection in human cells representative of the CNS [64, 65] and that HCoV-OC43 RNA could be detected for at least a year in the CNS of infected mice that survived the virus-induced acute encephalitis [71]. Therefore, an apparently innocuous human respiratory pathogen may persist in the human CNS as a component of what is proposed to be a «viral flora» of the brain, like HSV in a large proportion of the population. It would therefore be possible that such a persistent infection may become a factor or cofactor of neuropathogenesis in genetically or otherwise predisposed individuals.

Human coronaviruses were first isolated in the mid-1960s from patients with upper respiratory tract disease [72]. Until the end of the twentieth century, only two serological groups, represented by strains OC43 and 229E, were known to infect humans and they were recognized as respiratory pathogens responsible for up to 30 % of common colds [72]. Over the last 10 years, SARS has generated renewed interest in coronaviruses that led to the discovery of new coronaviruses that can infect humans: SARS-CoV [73, 74], HCoV-NL63 [75], HCoV-HKU1 [76] and MERS-CoV [77]. Among these six coronaviruses, at least HCoV-229E and HCoV-OC43, as well as SARS-CoV, possess neuroinvasive properties as viral RNA [63] or infectious virus [67, 68] can be detected in human brains.

To our knowledge, there exist no reports on the detection of HCoV-HKU1, HCoV-NL63, and MERS-CoV in the human CNS. On the other hand, neurological symptoms have been described in association with both HCoV-HKU1 and HCoV-NL63 [78] and a recent report, which evaluated MERS-CoV cell tropism, suggest that, among several cell lines representative of different tissues and organs, this virus seems to be able to infect the neuron-committed human cell line NT2 [79].

Possible Mechanisms of Neuroinvasiveness

Viruses may enter the CNS through two distinct routes: hematogenous dissemination or neuronal retrograde dissemination. Hematogenous spread involves the presence of a given virus in the bloodstream and retrograde viral spread toward the CNS occurs when a given virus infects neurons in the periphery and uses the transport machinery within those cells to gain access to the CNS [80].

In order to be neuroinvasive, viruses such as HCoV-229E, HCoV-OC43, and SARS-CoV may use both entry routes from the periphery. The hematogenous route involves the presence of a given virus in the blood, where it can either remain free for a period of time before it can infect the endothelial cells of the blood–brain-barrier (BBB), or infect leukocytes that will become some sort of viral reservoir for dissemination to other sites. Both situations occur during HIV infection of the CNS. Indeed, HIV-infected leukocytes that migrate through the BBB (called the Trojan horse [81]), is one of the route of spread, and direct infection of endothelial cells of the BBB is also possible even though the viral replication is at a low level [82].

Infection of human monocytes/macrophages by HCoV-229E and HCoV-OC43 was reported [83, 84] and infection by HCoV-229E of murine dentritic cells expressing the human aminopeptidase N [85] suggests that human coronaviruses may use these cells to disseminate to other tissues, including the CNS, where they could be associated with other type of pathologies. SARS-CoV was also shown to be able to infect human monocytes/macrophages [68, 86]. Moreover, monocyte-derived dendritic cells are also susceptible to infection by SARS-CoV [87].

Human primary monocytes are activated following infection by HCoV-229E [83]. Since they eventually become macrophages as they invade tissues, this activation suggests that HCoV-229E-infected monocytes would serve to facilitate their passage toward other tissues including the CNS, especially in immune-compromised individuals, as this was observed for murine cytomegalovirus (MCMV) [88]. The fact that HCoV-229E could only infect partially immune-compromised transgenic mice [89] suggests that HCoV-229E could take advantage of an immune-suppressed environment and disseminate to the CNS within susceptible individuals. The establishment of a persistent infection in a human leukocytic cell line [83] is also consistent with the possibility that monocytes/macrophages serve as a reservoir and vector for this neuroinvasive HCoV [63]. The SARS-CoV also infects monocytes/macrophages [68, 86] and dendritic cells, in which it modulates innate immunity [87]. These cells could also serve as a reservoir for the virus to reach and maintain itself in the CNS. Our results indicate that HCoV could also infect human endothelial cells of the BBB in culture (unpublished data) and it has been speculated that SARS-CoV could do the same after viremia [90]. Therefore, the neuroinvasive HCoV could use the hematogenous route to penetrate into the CNS.

The second form of any viral spread toward CNS is through neuronal dissemination, where a given virus infects neurons in periphery and uses the machinery of active transport within those cells in order to gain access to the CNS [80]. After an intranasal infection, both HCoV-OC43 [91] and SARS-CoV [92] were shown to infect the lungs in mice and to be neuroinvasive as HCoV-OC43 [93, 94] and SARS-CoV [95] were detected in the CNS of susceptible mice. Therefore, these two coronaviruses may use both the hematogenous and transneuronal route toward the CNS.

Furthermore, as shown in Fig. 1, once in the brain, HCoV-OC43 can disseminate from the olfactory bulb to the cortex and we have previously shown that it could also reach the medulla, while the cerebellum remained almost uninfected. The hippocampus represents another specific structure infected by HCoV-OC43 in the brain (Fig. 1) and once in this region of the brain, the virus appears to spread by a transneuronal route before it eventually reaches the spinal cord [48].

Mechanisms of HCoV-Induced Neurodegeneration and Programmed Cell Death: Possible Associated Neuropathologies

Neuroinvasive viruses can damage the CNS as a result of misdirected host immune responses (virus-induced neuroimmunopathology) and/or viral replication, which directly induces damage to CNS cells (virus-induced neuropathology). In acute encephalitis, viral replication occurs in the brain tissue itself, possibly causing destructive lesions of the gray matter [47]. As mentioned above (Section "Neuroinvasive and Neurotropic Viruses: Associated Neuropathologies"), chronic human neurological diseases may also be linked to viral infection. However, in several cases of these chronic diseases, it is difficult to ascertain a role for any given virus, in part due to the difficulty of establishing the time at which these viruses become involved. Also, the four Koch's postulates for disease induction dictate whether a particular infectious agent causes a specific disease [96].

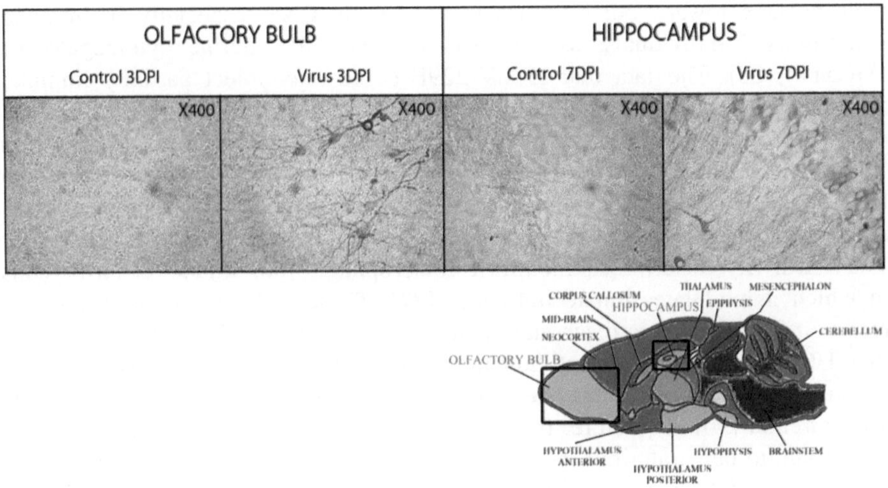

Fig. 1 Illustration of the transneuronal route used by HCoV-OC43 for neuroinvasion and dissemination into the central nervous system. The *left panel* shows the olfactory bulb area, either mock-infected (control) or HCoV-OC43-infected (virus) at 3 days postinfection (DPI). The *right panel* shows the hippocampus, either mock-infected (control) or HCoV-OC43-infected (virus) at 7 days postinfection (DPI). In both regions of the brain, neurons are the target of infection. Magnification is 400X

However, several viral infections, especially slow viral infections related to diseases that are rare manifestations of an infection, represent situations where Koch's postulate should be modified to better adapt to the situation [97, 98]. A series of new criteria, adapted from Sir Austin Bradford Hill's criteria for causation [98], has been elaborated by Giovannoni and collaborators [99] and should replace Koch's postulates when one wants to evaluate the relevance of any given virus in relation to MS etiology [99] or any other long-term human neurological diseases potentially related to a viral infection as well, including infection by human coronaviruses.

Possible Coronaviral-Induced Neuroimmunopathology

The presence of HCoV-229E and HCoV-OC43 was detected in various neurological diseases in humans, including PD and MS [63] and ADEM [100].

Multiple sclerosis truly represents a human neurological disease where an infectious agent or agents may play a triggering role, with viruses the most likely culprit in genetically predisposed individuals [101]. There is a presumption that several neurotropic viruses could be involved in MS pathogenesis but that they may do so through similar direct and/or indirect mechanisms [102–105]. However, research has not yet led to a direct link to any specific virus or other microbes with MS. Association of coronaviruses with MS was suggested in numerous reports that

Table 1 Detection of viral RNA in brain samples and of cross-reactive T-cell clones between myelin and HCoV antigens in control and MS patients

	Gender	HCoV-OC43		HCOV-229E		Both
Normal controls	Male	4(19)		9(19)		4(19)
	Female	1(5)		2(5)		0(5)
AD, PD, ALS and OND	Male	2(13)		6(13)		2(13)
	Female	1(13)		3(13)		0(13)
Multiple Sclerosis	Male	7(20)		11 (20)		5(20)
	Female	7(19)		9(19)		6(19)
T cell clones	*Monospecific*		*Cross reactive*			
	HCoV	Myelin	229E	OC43	MBP	PLP
Normal controls n = 6	28	7	0	0	0	0
Multiple Sclerosis n–32	114	31	4	2	2	2

Adapted from Arbour et al. [63] and Boucher et al. [108]
Numbers in the *upper portion* indicate the number of individuals positive for viral RNA and numbers in *parenthesis* indicate the total number of individuals tested
Numbers in the *lower portion* indicate the number of T-cell clones obtained. Monospecific describes clones that react against a single antigen and cross reactive describes clones that react both with HCoV and myelin antigens

are reviewed elsewhere [48]. One of these reports demonstrated a significant association of colds with MS exacerbations and a significant association of HCoV-229E infection in MS patients [106] and another report on the association of viral infections and MS [107] commented that seasonal HCoV infection patterns do fit the observed occurrence of MS exacerbations. Furthermore, the case of these human coronaviruses in the CNS may represent a new example where the traditional Koch's postulates should be replaced by the previously cited adapted Hill's criteria [99].

More than a decade ago, we experimentally confirmed that HCoV-OC43 and HCoV-229E were naturally neuroinvasive in humans. Although viruses were also detected in some control brains, there was a significantly higher prevalence of HCoV-OC43 in brains of MS patients [63]. Moreover, these data, in association with the observation that autoreactive T-cells recognized both viral and myelin antigens in MS patients but not in controls [108, 109] during infection by HCoV-OC43 and HCoV-229E, suggest that the immune response may participate in the induction or exacerbation of neuropathologies such as MS in genetically or otherwise susceptible individuals (data summarized in Table 1). Furthermore, even though the use of the immunosuppressive drug cyclosporin A in HCoV-OC43-infected mice resulted in a faster onset of encephalitis, suggesting a role for T-cells in viral clearance and survival with no related immunopathology [71], it was shown that in recombination activation gene (RAG) knock-out mice, HCoV-OC43-induced encephalitis could be partially mediated by the T-cell response to infection [93]. The participation of different types of T-cells has been shown to play a significant role in the demyelinating neurological disease induced by the

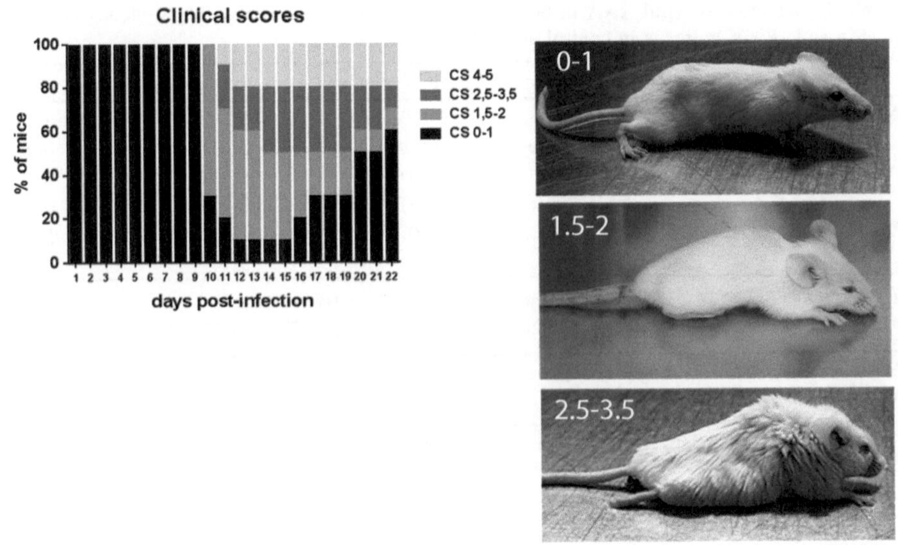

Fig. 2 Evaluation of the clinical scores (CS) related to motor dysfunctions (paralysis) of HCoV-OC43 infected mice. The *left panel* shows the percentage of mice observed at each of the different degree of motor dysfunctions over the course of infection, from 0 to 21 days postinfection. The CS were established according to a scale based on the recognized experimental allergic encephalitis (EAE) model (*0–1* normal mouse with no clinical signs; *1.5–2* partial hind-limb paralysis, with a walk close to ground level; *2.5–3.5* complete hind-limb paralysis, and *4–5* moribund state or death. The *right panels* are a representation of mice at each stage of paralysis

murine CoV, in particular for strain MHV-JHM [110], which represents the murine counterpart of HCoV-OC43.

More recently, making use of another mouse model, we showed that HCoV-OC43 induced immune cell infiltration and cytokine production in mouse CNS and that this response was significantly higher after infection by variants which harbor point mutations in the viral surface glycoprotein (S) [111]. Importantly, these S point mutations were acquired after persistent infections of human neural cell lines [112]. Moreover, we also showed that infection by the S mutants is linked to glutamate excitotoxicity [113]. Therefore, as this increase in cytokine production may induce damages to the neurons [114] that can be associated with problems in glutamate homeostasis, which in the end may create glutamate excitotoxicity [115], it can contribute to neuronal degeneration associated with hind-limb paralysis, as illustrated in Fig. 2, and possible demyelination [111, 113]. The outcome of the observed degeneration of neurons may eventually lead to the death of these essential cells.

As previously mentioned, infection of neurons by itself may also participate in the process of cell death by directly generating a cytotoxic insult related to viral replication and/or to the induction of different cell death pathways.

Direct Viral-Induced Neuropathology

When present in the murine CNS, HCoV-OC43 infects neurons in several different regions of the brain [48] (Fig. 1) before reaching the spinal cord. Infection of these essential cells induces their degeneration [111, 113], as observed by aberrant state of neurofilament phosphorylation, a situation that often leads to cell death and that could be directly induced by viral replication. Furthermore, using two model cell lines of differentiated human neurons, we demonstrated that programmed cell death (PCD) was induced after HCoV-OC43 infection [116, 117] and that the inhibition of viral replication was also in direct correlation with increased cell survival, suggesting that infection and production of new infectious viruses directly participate in the process of degeneration and eventual death of neurons. Our results indicate that the underlying mechanisms appear to involve different cellular factors and pathways, including caspase-independent apoptosis, parthanatos, and necroptosis, three forms of programmed cell death (PCD), which are reviewed elsewhere by the Nomenclature Committee on Cell Death (NCCD) [118]. These cell death pathways can act separately but may also interact in response to a stimulus (including a viral infection), as they share some of the cellular factors involved in the overall process that leads to cell death and that often converges toward mitochondria [118]. Figure 3 is a tentative representation of the various pathways and cellular factors involved during HCoV-OC43-induced PCD of infected neurons. It is based on our data [111, 113, 116, 117] and on the scientific literature that describes some molecular pathways (parthanatos, necroptosis and apoptosis) and cellular factors, including calcium overload, endoplasmic reticulum (ER) stress, excitotoxicity, poly(ADP-ribose) polymerase (PARP), calpain and oxidative stress related to the formation of reactive oxygen species (ROS) involved in mitochondrial dysfunction, and eventual neurodegeneration and neuronal cell death [118–120]. Virus–cell interaction is always important in the regulation of cell response to infection. For HCoV-OC43, we clearly showed that the viral S glycoprotein is an important factor of neurovirulence and neurodegeneration of infected cells [111, 113, 116]. This was similarly shown for coronavirus strains MHV-A59 and MHV-JHM, which represent subspecies of the MuCoV species, the murine counterpart of HCoV-OC43 as they are both members of the Betacoronavirus genus. Indeed, several reports and reviews have, over the years, described that S protein of this neuroinvasive and neurotropic murine coronavirus is a major factor associated with neurovirulence during encephalitis and the eventual demyelinating disease in susceptible mice [121, 122]. Our more recent data also demonstrated that the HE protein is an important factor for the production of infectious HCoV-OC43, suggesting an attenuation of the eventual spread of viruses deficient in fully active HE protein into the CNS [123]. Therefore, as the infection of neuronal cells apparently directly participate in the induction of neuronal death, by abrogating the production of infectious virus, the HE protein of HCoV-OC43 could play a role in neurovirulence of HCoV-OC43, like it does for MHV [124].

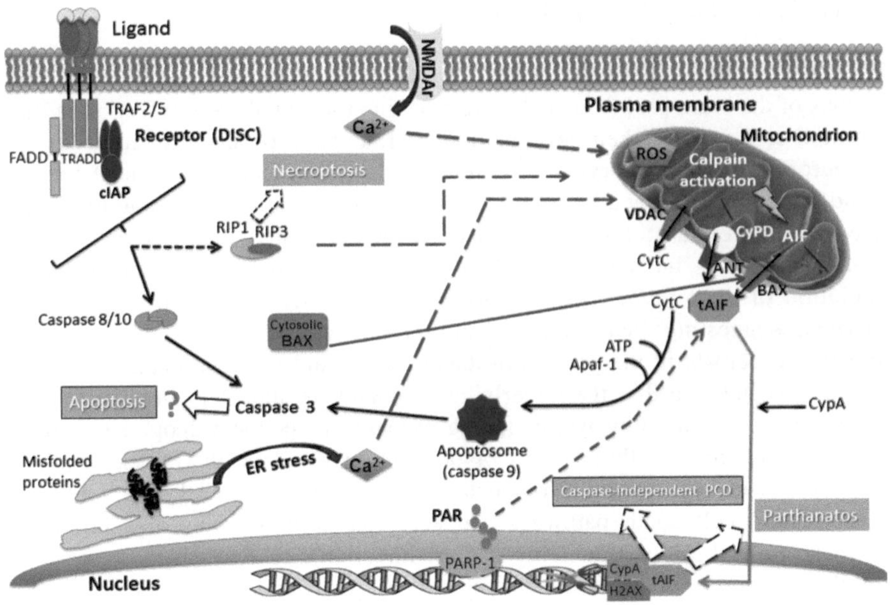

Fig. 3 Pathways of neuronal degeneration and cell death induced by HCoV–OC43 infection. Several cellular factors that regulate various mechanisms are activated in response to infection, which leads to programmed cell death (PCD). (1) Hallmarks of apoptosis, including relocalization of the activated pro-apoptotic protein BAX (Bcl-2 associated protein X) from the cytosol to the mitochondrial membrane, cytochrome C release from mitochondria toward the cytosol, DNA fragmentation, and activation of caspases -3 and -9, are observed during infection of human neurons. However, using a pan-caspase inhibitor (Z-VAD-fmk), cell death is not abrogated after infection, suggesting a caspase-independent type of apoptosis. (2) Relocalization of the mitochondrial protein AIF (apoptosis-inducing factor) toward the nucleus (tAIF) is observed after infection and participates in DNA fragmentation in conjunction with CypA (cyclophilin A) and histone H2AX. The AIF is known to be activated during caspase-independent apoptosis. However, AIF is also involved in Parthanatos, another form of PCD potentially associated with neurodegeneration. As they are synthesized by the poly(ADP-ribose) polymerase (PARP) during a neuronal stress, including during HCoV–OC43 infection, polymers of ADP-ribose (PAR) may relocalize toward mitochondria and participate in the activation and relocalization of AIF toward the cytosol before it reaches the nucleus. Cyclophilin D (CypD) inhibition decreases AIF release from mitochondria and abrogates cell death induced by infection. (3) AIF release from mitochondria may be induced through its truncation (tAIF) by activated calpain, which is usually activated by a rise in the mitochondrial calcium concentration. (4) This increase in calcium concentration may be linked with either an important entry from the extracellular milieu (for instance during excitotoxicity) or with a release of calcium from the endoplasmic reticulum (ER) following induction of ER stress. Both situations are probably taking place after infection of neurons by HCoV-OC43. The increase in calcium concentration in mitochondria may also induce production of reactive oxygen species (ROS) that can be harmful for mitochondria and hence neurons. (5) The presence during infection of an inhibitor (Nec-1) of the receptor interacting protein kinase-1 (RIP-1), significantly increases cell survival and partially abrogates viral replication, suggesting that necroptosis, a third form of PCD which involves RIP-1 and RIP-3 downstream of the tumor necrosing factor (TNF) receptor family (in the form of the death-inducing signaling complex (DISC)), may play a role in HCoV-OC43-induced neuronal death. Solid arrows indicate experimental data and dashed arrows represent possible pathways based on the current literature (see text for details)

As mentioned above, SARS-CoV is also neuroinvasive and neurotropic in humans [67, 68] and it could therefore be associated with the development of a neurological disease. Furthermore, the involvement of SARS-CoV in CNS infections was underscored by the findings that made use of transgenic mouse models expressing the human angiotensin-converting enzyme-2 (the cellular receptor used by SARS-CoV to infect susceptible cells). Indeed, using these mice, it was shown that SARS-CoV could invade the CNS after an intranasal infection primarily through the olfactory bulb [95] or even after an intraperitoneal infection [125], with concomitant neuronal loss [95, 125]; a phenomenon that can eventually lead to neurological problems.

Conclusions and Significance

The presence of coronaviruses in the human central nervous system is now a recognized fact as they appear to be part of a viral flora of the brain, with potential neuropathological consequences in genetically or otherwise susceptible individuals, with or without additional environmental insults. Knowledge of mechanisms and consequences of virus interactions with the nervous system is essential to better understand potentially pathological consequences and design intervention strategies that are appropriate to encephalitis or exacerbations of other types of neurological diseases for which a given virus is involved. In that regard, Hill's criteria adapted by Giovannoni and collaborators may represent a highly relevant tool to evaluate the relevance of human coronaviruses as a factor which will influence the development and/or exacerbation of a chronic human neurological disease potentially related to a viral infection. Therefore, collecting new data will be instrumental to our understanding of how the ubiquitous human coronaviruses, given the proper susceptibility conditions and proper virus evolution and infection conditions, could participate in the induction or exacerbation of human neuropathologies.

References

1. Vareille M, Kieninger E, Edwards MR, Regamey N (2011) The airway epithelium: soldier in the fight against respiratory viruses. Clin Microbiol Rev 24(1):210–229. doi:10.1128/CMR.00014-10
2. Jartti T, Jartti L, Ruuskanen O, Soderlund-Venermo M (2012) New respiratory viral infections. Curr Opin Pulm Med 18(3):271–278. doi:10.1097/MCP.0b013e328351f8d4
3. Ison MG, Hayden FG (2002) Viral infections in immunocompromised patients: what's new with respiratory viruses? Curr Opin Infect Dis 15(4):355–367
4. Cesario TC (2012) Viruses associated with pneumonia in adults. Clin Infect Dis 55(1):107–113. doi:10.1093/cid/cis297

5. Sloots TP, Whiley DM, Lambert SB, Nissen MD (2008) Emerging respiratory agents: new viruses for old diseases? J Clin Virol 42(3):233–243. doi:10.1016/j.jcv.2008.03.002
6. Brouard J, Vabret A, Nimal-Cuvillon D, Bach N, Bessiere A, Arion A, Freymuth F (2007) Epidemiology of acute upper and lower respiratory tract infections in children. Rev Prat 57(16):1759–1766
7. McGavern DB, Kang SS (2011) Illuminating viral infections in the nervous system. Nat Rev Immunol 11(5):318–329. doi:10.1038/nri2971
8. Giraudon P, Bernard A (2010) Inflammation in neuroviral diseases. J Neural Trans 117(8):899–906. doi:10.1007/s00702-010-0402-y
9. Whitley RJ, Gnann JW (2002) Viral encephalitis: familiar infections and emerging pathogens. Lancet 359(9305):507–513. doi:10.1016/S0140-6736(02)07681-X
10. Hankins DG, Rosekrans JA (2004) Overview, prevention, and treatment of rabies. Mayo Clin Proc 79(5):671–676
11. Aurelian L (2005) HSV-induced apoptosis in herpes encephalitis. Curr Top Microbiol Immunol 289:79–111
12. Mackenzie JS, Gubler DJ, Petersen LR (2004) Emerging flaviviruses: the spread and resurgence of Japanese encephalitis, West Nile and dengue viruses. Nat Med 10(12 suppl):S98–S109. doi:10.1038/nm1144
13. Mattson MP, Haughey NJ, Nath A (2005) Cell death in HIV dementia. Cell Death Differ 12(1):893–904. doi:10.1038/sj.cdd.4401577
14. Nath A, Berger J (2004) HIV Dementia. Curr Treat Opt Neurol 6(2):139–151
15. Gordon J, Gallia GL, Del Valle L, Amini S, Khalili K (2000) Human polyomavirus JCV and expression of myelin genes. J Neurovirol 6(suppl 2):S92–S97
16. Weissert R (2011) Progressive multifocal leukoencephalopathy. J Neuroimmunol 231(1–2):73–77. doi:10.1016/j.jneuroim.2010.09.021
17. Rima BK, Duprex WP (2005) Molecular mechanisms of measles virus persistence. Virus Res 111(2):132–147. doi:10.1016/j.virusres.2005.04.005
18. Kaplan JE, Osame M, Kubota H, Igata A, Nishitani H, Maeda Y, Khabbaz RF, Janssen RS (1990) The risk of development of HTLV-I-associated myelopathy/tropical spastic paraparesis among persons infected with HTLV-I. J Acquir Immune Defic Syndr 3(11):1096–1101
19. Itzhaki RF, Wozniak MA, Appelt DM, Balin BJ (2004) Infiltration of the brain by pathogens causes Alzheimer's disease. Neurobiol Aging 25(5):619–627. doi:10.1016/j.neurobiolaging.2003.12.021
20. Stensballe LG, Devasundaram JK, Simoes EA (2003) Respiratory syncytial virus epidemics: the ups and downs of a seasonal virus. Pediatr Infect Dis J 22(2):S21–S32. doi:10.1097/01.inf.0000053882.70365.c9
21. Kawashima H, Ioi H, Ushio M, Yamanaka G, Matsumoto S, Nakayama T (2009) Cerebrospinal fluid analysis in children with seizures from respiratory syncytial virus infection. Scand J Infect Dis 41(3):228–231. doi:10.1080/00365540802669543
22. Zlateva KT, Van Ranst M (2004) Detection of subgroup B respiratory syncytial virus in the cerebrospinal fluid of a patient with respiratory syncytial virus pneumonia. Pediatr Infect Dis J 23(11):1065–1066
23. Morichi S, Kawashima H, Ioi H, Yamanaka G, Kashiwagi Y, Hoshika A, Nakayama T, Watanabe Y (2011) Classification of acute encephalopathy in respiratory syncytial virus infection. J Infect Chemother 17(6):776–781. doi:10.1007/s10156-011-0259-5
24. Millichap JJ, Wainwright MS (2009) Neurological complications of respiratory syncytial virus infection: case series and review of literature. J Child Neurol 24(12):1499–1503. doi:10.1177/0883073808331362
25. Espinoza JA, Bohmwald K, Cespedes PF, Gomez RS, Riquelme SA, Cortes CM, Valenzuela JA, Sandoval RA, Pancetti FC, Bueno SM, Riedel CA, Kalergis AM (2013) Impaired learning resulting from respiratory syncytial virus infection. Proc Natl Acad Sci USA 110(22):9112–9117. doi:10.1073/pnas.1217508110

26. O'Donnell LA, Rall GF (2010) Blue moon neurovirology: the merits of studying rare CNS diseases of viral origin. J Neuroimmune Pharmacol 5(3):443–455. doi:10.1007/s11481-010-9200-4
27. Wilson MR, Ludlow M, Duprex WP (2013) Human paramyxoviruses and infections of the central nervous system. In: Singh SK, Ruzek D (eds) Neuroviral infections. RNA viruses and retroviruses. CRC Press/Taylor and Francis, Boca Raton, pp 341–372
28. Escaffre O, Borisevich V, Rockx B (2013) Pathogenesis of Hendra and Nipah virus infection in humans. J Infect Dev Ctries 7(4):308–311. doi:10.3855/jidc.3648
29. Wong KT (2010) Emerging epidemic viral encephalitides with a special focus on henipaviruses. Acta Neuropathol 120(3):317–325. doi:10.1007/s00401-010-0720-z
30. Munster VJ, Prescott JB, Bushmaker T, Long D, Rosenke R, Thomas T, Scott D, Fischer ER, Feldmann H, de Wit E (2012) Rapid Nipah virus entry into the central nervous system of hamsters via the olfactory route. Sci Rep 2:736. doi:10.1038/srep00736
31. Zeng J, Wang G, Kang-Sheng L (2013) Influenza virus and CNS infections. In: Singh SK, Ruzek D (eds) Neuroviral Infections. RNA viruses and retroviruses. CRC Press/Taylor and Francis, Boca Raton, pp 325–339
32. Kuiken T, Riteau B, Fouchier RA, Rimmelzwaan GF (2012) Pathogenesis of influenza virus infections: the good, the bad and the ugly. Curr Opin Virol 2(3):276–286. doi:10.1016/j.coviro.2012.02.013
33. Matsuzaki Y, Katsushima N, Nagai Y, Shoji M, Itagaki T, Sakamoto M, Kitaoka S, Mizuta K, Nishimura H (2006) Clinical features of influenza C virus infection in children. J Infect Dis 193(9):1229–1235. doi:10.1086/502973
34. Gouarin S, Vabret A, Dina J, Petitjean J, Brouard J, Cuvillon-Nimal D, Freymuth F (2008) Study of influenza C virus infection in France. J Med Virol 80(8):1441–1446. doi:10.1002/jmv.21218
35. Munier S, Moisy D, Marc D, Naffakh N (2010) Interspecies transmission, adaptation to humans and pathogenicity of animal influenza viruses. Pathol Biol 58(2):e59–e68. doi:10.1016/j.patbio.2010.01.012
36. Nicholson KG, Wood JM, Zambon M (2003) Influenza. Lancet 362(9397):1733–1745. doi:10.1016/S0140-6736(03)14854-4
37. Jang H, Boltz DA, Webster RG, Smeyne RJ (2009) Viral parkinsonism. Biochim Biophys Acta 1792(7):714–721. doi:10.1016/j.bbadis.2008.08.001
38. Wang GF, Li W, Li K (2010) Acute encephalopathy and encephalitis caused by influenza virus infection. Curr Opin Neurol 23(3):305–311
39. Millichap JG, Millichap JJ (2006) Role of viral infections in the etiology of febrile seizures. Pediatr Neurol 35(3):165–172. doi:10.1016/j.pediatrneurol.2006.06.004
40. Toovey S (2008) Influenza-associated central nervous system dysfunction: a literature review. Travel Med Infect Dis 6(3):114–124. doi:10.1016/j.tmaid.2008.03.003
41. Ozkale Y, Erol I, Ozkale M, Demir S, Alehan F (2012) Acute disseminated encephalomyelitis associated with influenza A H1N1 infection. Pediatr Neurol 47(1):62–64. doi:10.1016/j.pediatrneurol.2012.03.019
42. Beraki S, Aronsson F, Karlsson H, Ogren SO, Kristensson K (2005) Influenza A virus infection causes alterations in expression of synaptic regulatory genes combined with changes in cognitive and emotional behaviors in mice. Mol Psychiatry 10(3):299–308. doi:10.1038/sj.mp.4001545
43. Jurgens HA, Amancherla K, Johnson RW (2012) Influenza infection induces neuroinflammation, alters hippocampal neuron morphology, and impairs cognition in adult mice. J Neurosci 32(12):3958–3968. doi:10.1523/JNEUROSCI.6389-11.2012
44. Vabret A, Dina J, Brison E, Brouard J, Freymuth F (2009) Human coronaviruses. Pathol Biol 57(2):149–160. doi:10.1016/j.patbio.2008.02.018
45. Buchmeier MJ, Lane TE (1999) Viral-induced neurodegenerative disease. Curr Opin Microbiol 2 (4):398–402. doi:mc2416 [pii]
46. Cavanagh D (2005) Coronaviruses in poultry and other birds. Avian Pathol 34(6):439–448. doi:10.1080/03079450500367682

47. Talbot PJ, Desforges, M, Brison, E, Jacomy, H (2011) Coronaviruses as Encephalitis-inducing infectious agents. In: Tkachev S (ed) Non-flavirus Encephalitis. In-Tech, pp 185–202
48. Desforges M, Favreau DJ, Brison E, Desjardins J, Meessen-Pinard M, Jacomy H, Talbot PJ (2013) Human coronaviruses. Respiratory pathogens revisited as infectious neuroinvasive, neurotropic, and neurovirulent agents. In: Singh SK, Ruzek D (eds) Neuroviral infections. RNA viruses and retroviruses. CRC Press/Taylor and Francis, Boca Raton, pp 93–121
49. de Groot RJ, Baker SC, Baric R, Enjuanes L, Gorbalenya AE, Holmes KV, Perlman S, Poon L, Rottier PJM, Talbot PJ, Woo PCY, Ziebuhr J (eds) (2012) Family coronaviridae. Virus taxonomy: ninth report of the international commitee on taxonomy of viruses. Elsevier, New York
50. de Groot RJ, Baker SC, Baric RS, Brown CS, Drosten C, Enjuanes L, Fouchier RA, Galiano M, Gorbalenya AE, Memish Z, Perlman S, Poon LL, Snijder EJ, Stephens GM, Woo PC, Zaki AM, Zambon M, Ziebuhr J (2013) Middle East respiratory syndrome coronavirus (MERS-CoV); Announcement of the Coronavirus Study Group. J Virol. doi:10.1128/JVI.01244-13
51. Saif LJ (2008) Coronaviruses of domestic livestock and poultry: interspecies transmission, pathogenesis, and immunity. In: Perlman S, Gallagher T, Snijder EJ (eds) Nidoviruses. ASM Press, Washington D.C, pp 279–298
52. Foley JE, Lapointe JM, Koblik P, Poland A, Pedersen NC (1998) Diagnostic features of clinical neurologic feline infectious peritonitis. J Vet Intern Med 12(6):415–423
53. Kline K, Joseph R, Averill DAJ (1994) Feline infectious peritonitis with neurological involvement: clinical and pathological findings in 24 cats. J Am Anim Hosp Assoc 30:111–118
54. Foley JE, Rand C, Leutenegger C (2003) Inflammation and changes in cytokine levels in neurological feline infectious peritonitis. J Feline Med Surg 5(6):313–322
55. Slauson DO, Finn JP (1972) Meningoencephalitis and panophthalmitis in feline infectious peritonitis. J Am Vet Med Assoc 160(5):729–734
56. Legendre AM, Whitenack DL (1975) Feline infectious peritonitis with spinal cord involvement in two cats. J Am Vet Med Assoc 167(10):31–32
57. Greig AS, Mitchell D, Corner AH, Bannister GL, Meads EB, Julian RJ (1962) A hemagglutinating virus producing encephalomyelitis in baby pigs. Can J Comp Med Vet Sci 26(3):49–56
58. Alexander TJ (1962) Viral encephalomyelitis of swine in Ontario–experimental and natural transmission. Am J Vet Res 23:756–762
59. Andries K, Pensaert MB (1980) Virus isolated and immunofluorescence in different organs of pigs infected with hemagglutinating encephalomyelitis virus. Am J Vet Res 41(2):215–218
60. Hirano N, Nomura R, Tawara T, Tohyama K (2004) Neurotropism of swine haemagglutinating encephalomyelitis virus (coronavirus) in mice depending upon host age and route of infection. J Comp Pathol 130(1):58–65
61. Weiss SR, Leibowitz JL (2011) Coronavirus pathogenesis. Advances in virus research, vol 81. Elsevier, Boston, pp 85–164
62. Brian DA, Baric RS (2005) Coronavirus genome structure and replication. Curr Top Microbiol Immunol 287:1–30
63. Arbour N, Day R, Newcombe J, Talbot PJ (2000) Neuroinvasion by human respiratory coronaviruses. J Virol 74(19):8913–8921
64. Arbour N, Cote G, Lachance C, Tardieu M, Cashman NR, Talbot PJ (1999) Acute and persistent infection of human neural cell lines by human coronavirus OC43. J Virol 73(4):3338–3350
65. Arbour N, Ekande S, Cote G, Lachance C, Chagnon F, Tardieu M, Cashman NR, Talbot PJ (1999) Persistent infection of human oligodendrocytic and neuroglial cell lines by human coronavirus 229E. J Virol 73(4):3326–3337

66. Bonavia A, Arbour N, Yong VW, Talbot PJ (1997) Infection of primary cultures of human neural cells by human coronaviruses 229E and OC43. J Virol 71(1):800–806
67. Xu J, Zhong S, Liu J, Li L, Li Y, Wu X, Li Z, Deng P, Zhang J, Zhong N, Ding Y, Jiang Y (2005) Detection of severe acute respiratory syndrome coronavirus in the brain: potential role of the chemokine mig in pathogenesis. Clin Infect Dis 41(8):1089–1096. doi:10.1086/444461
68. Gu J, Gong E, Zhang B, Zheng J, Gao Z, Zhong Y, Zou W, Zhan J, Wang S, Xie Z, Zhuang H, Wu B, Zhong H, Shao H, Fang W, Gao D, Pei F, Li X, He Z, Xu D, Shi X, Anderson VM, Leong AS (2005) Multiple organ infection and the pathogenesis of SARS. J Exp Med 202(3):415–424. doi:10.1084/jem.20050828
69. Talbot PJ, Jacomy H, Desforges M (2008) Pathogenesis of human coronaviruses other than severe acute respiratory syndrome coronavirus. In: Perlman S, Gallagher T, Snijder EJ (eds) Nidoviruses. ASM Press, Washington D.C, pp 313–324
70. http://www.who.int/csr/don/2013_08_30/en/index.html
71. Jacomy H, Fragoso G, Almazan G, Mushynski WE, Talbot PJ (2006) Human coronavirus OC43 infection induces chronic encephalitis leading to disabilities in BALB/C mice. Virology 349(2):335–346. doi:10.1016/j.virol.2006.01.049
72. Myint SH (1995) Human coronavirus infections. In: Siddell SG (ed) The coronaviridae. Plenum Press, New York, pp 389–401
73. Drosten C, Gunther S, Preiser W, van der Werf S, Brodt HR, Becker S, Rabenau H, Panning M, Kolesnikova L, Fouchier RA, Berger A, Burguiere AM, Cinatl J, Eickmann M, Escriou N, Grywna K, Kramme S, Manuguerra JC, Muller S, Rickerts V, Sturmer M, Vieth S, Klenk HD, Osterhaus AD, Schmitz H, Doerr HW (2003) Identification of a novel coronavirus in patients with severe acute respiratory syndrome. N Eng J Med 348(20):1967–1976. doi:10.1056/NEJMoa030747
74. Fouchier RA, Kuiken T, Schutten M, van Amerongen G, van Doornum GJ, van den Hoogen BG, Peiris M, Lim W, Stohr K, Osterhaus AD (2003) Aetiology: Koch's postulates fulfilled for SARS virus. Nature 423(6937):240. doi:10.1038/423240a
75. van der Hoek L, Pyrc K, Jebbink MF, Vermeulen-Oost W, Berkhout RJ, Wolthers KC, Wertheim-van Dillen PM, Kaandorp J, Spaargaren J, Berkhout B (2004) Identification of a new human coronavirus. Nat Med 10(4):368–373. doi:10.1038/nm1024
76. Woo PC, Lau SK, Chu CM, Chan KH, Tsoi HW, Huang Y, Wong BH, Poon RW, Cai JJ, Luk WK, Poon LL, Wong SS, Guan Y, Peiris JS, Yuen KY (2005) Characterization and complete genome sequence of a novel coronavirus, coronavirus HKU1, from patients with pneumonia. J Virol 79(2):884–895. doi:10.1128/JVI.79.2.884-895.2005
77. Zaki AM, van Boheemen S, Bestebroer TM, Osterhaus AD, Fouchier RA (2012) Isolation of a novel coronavirus from a man with pneumonia in Saudi Arabia. N Engl J Med 367(19):1814–1820. doi:10.1056/NEJMoa1211721
78. Severance EG, Dickerson FB, Viscidi RP, Bossis I, Stallings CR, Origoni AE, Sullens A, Yolken RH (2011) Coronavirus immunoreactivity in individuals with a recent onset of psychotic symptoms. Schizophr Bull 37(1):101–107. doi:10.1093/schbul/sbp052
79. Fuk-Woo Chan J, Chan KH, Choi GK, To KK, Tse H, Cai JP, Yeung ML, Cheng VC, Chen H, Che XY, Lau SK, Woo PC, Yuen KY (2013) Differential cell line susceptibility to the emerging novel human betacoronavirus 2c EMC/2012: implications for disease pathogenesis and clinical manifestation. J Infect Dis 207(11):1743–1752. doi:10.1093/infdis/jit123
80. Berth SH, Leopold PL, Morfini GN (2009) Virus-induced neuronal dysfunction and degeneration. Front Biosci 14:5239–5259
81. Kim WK, Corey S, Alvarez X, Williams K (2003) Monocyte/macrophage traffic in HIV and SIV encephalitis. J Leukoc Biol 74(5):650–656. doi:10.1189/jlb.0503207
82. Argyris EG, Acheampong E, Wang F, Huang J, Chen K, Mukhtar M, Zhang H (2007) The interferon-induced expression of APOBEC3G in human blood–brain barrier exerts a potent intrinsic immunity to block HIV-1 entry to central nervous system. Virology 367(2):440–451. doi:10.1016/j.virol.2007.06.010

83. Desforges M, Miletti TC, Gagnon M, Talbot PJ (2007) Activation of human monocytes after infection by human coronavirus 229E. Virus Res 130(1–2):228–240. doi:10.1016/j.virusres.2007.06.016
84. Collins AR (2002) In vitro detection of apoptosis in monocytes/macrophages infected with human coronavirus. Clin Diagn Lab Immunol 9(6):1392–1395
85. Wentworth DE, Tresnan DB, Turner BC, Lerman IR, Bullis B, Hemmila EM, Levis R, Shapiro LH, Holmes KV (2005) Cells of human aminopeptidase N (CD13) transgenic mice are infected by human coronavirus-229E in vitro, but not in vivo. Virology 335(2):185–197. doi:10.1016/j.virol.2005.02.023
86. Nicholls JM, Butany J, Poon LL, Chan KH, Beh SL, Poutanen S, Peiris JS, Wong M (2006) Time course and cellular localization of SARS-CoV nucleoprotein and RNA in lungs from fatal cases of SARS. PLoS Med 3(2):e27. doi:10.1371/journal.pmed.0030027
87. Spiegel M, Schneider K, Weber F, Weidmann M, Hufert FT (2006) Interaction of severe acute respiratory syndrome-associated coronavirus with dendritic cells. J Gen Virol 87(Pt 7):1953–1960. doi:10.1099/vir.0.81624-0
88. Reuter JD, Gomez DL, Wilson JH, Van Den Pol AN (2004) Systemic immune deficiency necessary for cytomegalovirus invasion of the mature brain. J Virol 78(3):1473–1487
89. Lassnig C, Sanchez CM, Egerbacher M, Walter I, Majer S, Kolbe T, Pallares P, Enjuanes L, Muller M (2005) Development of a transgenic mouse model susceptible to human coronavirus 229E. Proc Nat Acad Sci U S A 102(23):8275–8280. doi:10.1073/pnas.0408589102
90. Guo Y, Korteweg C, McNutt MA, Gu J (2008) Pathogenetic mechanisms of severe acute respiratory syndrome. Virus Res 133(1):4–12. doi:10.1016/j.virusres.2007.01.022
91. Jacomy H, Talbot PJ (2003) Vacuolating encephalitis in mice infected by human coronavirus OC43. Virology 315(1):20–33
92. McCray PB Jr, Pewe L, Wohlford-Lenane C, Hickey M, Manzel L, Shi L, Netland J, Jia HP, Halabi C, Sigmund CD, Meyerholz DK, Kirby P, Look DC, Perlman S (2007) Lethal infection of K18-hACE2 mice infected with severe acute respiratory syndrome coronavirus. J Virol 81(2):813–821. doi:10.1128/JVI.02012-06
93. Butler N, Pewe L, Trandem K, Perlman S (2006) Murine encephalitis caused by HCoV-OC43, a human coronavirus with broad species specificity, is partly immune-mediated. Virology 347(2):410–421. doi:10.1016/j.virol.2005.11.044
94. St-Jean JR, Jacomy H, Desforges M, Vabret A, Freymuth F, Talbot PJ (2004) Human respiratory coronavirus OC43: genetic stability and neuroinvasion. J Virol 78(16):8824–8834. doi:10.1128/JVI.78.16.8824-8834.2004
95. Netland J, Meyerholz DK, Moore S, Cassell M, Perlman S (2008) Severe acute respiratory syndrome coronavirus infection causes neuronal death in the absence of encephalitis in mice transgenic for human ACE2. J Virol 82(15):7264–7275. doi:10.1128/JVI.00737-08
96. Koch R (1942) The aetiology of tuberculosis (translation of Die Aetiologie der Tuberculose (1882). Dover Publications, New York
97. Fredericks DN, Relman DA (1996) Sequence-based identification of microbial pathogens: a reconsideration of Koch's postulates. Clin Microbiol Rev 9(1):18–33
98. Hill AB (1965) The environment and disease: association or causation? Proc R Soc Med 58:295–300
99. Giovannoni G, Cutter GR, Lunemann J, Martin R, Munz C, Sriram S, Steiner I, Hammerschlag MR, Gaydos CA (2006) Infectious causes of multiple sclerosis. Lancet Neurol 5(10):887–894. doi:10.1016/S1474-4422(06)70577-4
100. Yeh EA, Collins A, Cohen ME, Duffner PK, Faden H (2004) Detection of coronavirus in the central nervous system of a child with acute disseminated encephalomyelitis. Pediatrics 113(1 Pt 1):e73–e76
101. Kurtzke JF (1993) Epidemiologic evidence for multiple sclerosis as an infection. Clin Microbiol Rev 6(4):382–427
102. Cusick MF, Libbey JE, Fujinami RS (2013) Multiple sclerosis: autoimmunity and viruses. Curr Opin Rheumatol 25(4):496–501. doi:10.1097/BOR.0b013e328362004d

103. Kakalacheva K, Munz C (1812) Lunemann JD (2011) Viral triggers of multiple sclerosis. Biochim Biophys Acta 2:132–140. doi:10.1016/j.bbadis.2010.06.012
104. Gilden DH (2005) Infectious causes of multiple sclerosis. Lancet Neurol 4(3):195–202. doi:10.1016/S1474-4422(05)01017-3
105. Talbot PJ, Arnold D, Antel JP (2001) Virus-induced autoimmune reactions in the CNS. Curr Top Microbiol Immunol 253:247–271
106. Hovanec DL, Flanagan TD (1983) Detection of antibodies to human coronaviruses 229E and OC43 in the sera of multiple sclerosis patients and normal subjects. Infect Immun 41(1):426–429
107. Sibley WA, Bamford CR, Clark K (1985) Clinical viral infections and multiple sclerosis. Lancet 1(8441):1313–1315
108. Boucher A, Desforges M, Duquette P, Talbot PJ (2007) Long-term human coronavirus-myelin cross-reactive T-cell clones derived from multiple sclerosis patients. Clin Immunol 123(3):258–267. doi:10.1016/j.clim.2007.02.002
109. Talbot PJ, Paquette JS, Ciurli C, Antel JP, Ouellet F (1996) Myelin basic protein and human coronavirus 229E cross-reactive T-cells in multiple sclerosis. Ann Neurol 39(2):233–240. doi:10.1002/ana.410390213
110. Matthews AE, Weiss SR, Paterson Y (2002) Murine hepatitis virus—a model for virus-induced CNS demyelination. J Neurovirol 8(2):76–85. doi:10.1080/13550280290049534
111. Jacomy H, St-Jean JR, Brison E, Marceau G, Desforges M, Talbot PJ (2010) Mutations in the spike glycoprotein of human coronavirus OC43 modulate disease in BALB/c mice from encephalitis to flaccid paralysis and demyelination. J Neurovirol 16(4):279–293. doi:10.3109/13550284.2010.497806
112. St-Jean JR, Desforges M, Talbot PJ (2006) Genetic evolution of human coronavirus OC43 in neural cell culture. Adv Exp Med Biol 581:499–502. doi:10.1007/978-0-387-33012-9_88
113. Brison E, Jacomy H, Desforges M, Talbot PJ (2011) Glutamate excitotoxicity is involved in the induction of paralysis in mice after infection by a human coronavirus with a single point mutation in its spike protein. J Virol 85(23):12464–12473. doi:10.1128/JVI.05576-11
114. Amor S, Puentes F, Baker D, van der Valk P (2010) Inflammation in neurodegenerative diseases. Immunology 129(2):154–169. doi:10.1111/j.1365-2567.2009.03225.x
115. Carmen J, Rothstein JD, Kerr DA (2009) Tumor necrosis factor-alpha modulates glutamate transport in the CNS and is a critical determinant of outcome from viral encephalomyelitis. Brain Res 1263:143–154. doi:10.1016/j.brainres.2009.01.040
116. Favreau DJ, Desforges M, St-Jean JR, Talbot PJ (2009) A human coronavirus OC43 variant harboring persistence-associated mutations in the S glycoprotein differentially induces the unfolded protein response in human neurons as compared to wild-type virus. Virology 395(2):255–267. doi:10.1016/j.virol.2009.09.026
117. Favreau DJ, Desforges M, Talbot PJ (2011) Human coronavirus-induced neuronal programmed cell death is cyclophilin D-dependent and potentially caspase-dispensable. J Virol 86(1):81–93. doi:10.1128/JVI.06062-11
118. Galluzzi L, Vitale I, Abrams JM, Alnemri ES, Baehrecke EH, Blagosklonny MV, Dawson TM, Dawson VL, El-Deiry WS, Fulda S, Gottlieb E, Green DR, Hengartner MO, Kepp O, Knight RA, Kumar S, Lipton SA, Lu X, Madeo F, Malorni W, Mehlen P, Nunez G, Peter ME, Piacentini M, Rubinsztein DC, Shi Y, Simon HU, Vandenabeele P, White E, Yuan J, Zhivotovsky B, Melino G, Kroemer G (2012) Molecular definitions of cell death subroutines: recommendations of the nomenclature committee on cell death. Cell Death Differ 19(1):107–120. doi:10.1038/cdd.2011.96
119. Kaiser WJ, Upton JW, Mocarski ES (2013) Viral modulation of programmed necrosis. Curr Opin Virol. doi:10.1016/j.coviro.2013.05.019
120. Cali T, Ottolini D, Brini M (2011) Mitochondria, calcium, and endoplasmic reticulum stress in Parkinson's disease. BioFactors 37(3):228–240. doi:10.1002/biof.159
121. Bender SJ, Weiss SR (2010) Pathogenesis of murine coronavirus in the central nervous system. J Neuroimmune Pharmacol 5(3):336–354. doi:10.1007/s11481-010-9202-2

122. Hosking MP, Lane TE (2010) The pathogenesis of murine coronavirus infection of the central nervous system. Crit Rev Immunol 30(2):119–130
123. Desforges M, Desjardins J, Zhang C, Talbot PJ (2013) The acetyl-esterase activity of the hemagglutinin-esterase protein of human coronavirus OC43 strongly enhances the production of infectious virus. J Virol 87(6):3097–3107. doi:10.1128/JVI.02699-1
124. Kazi L, Lissenberg A, Watson R, de Groot RJ, Weiss SR (2005) Expression of hemagglutinin esterase protein from recombinant mouse hepatitis virus enhances neurovirulence. J Virol 79(24):15064–15073. doi:10.1128/JVI.79.24.15064-15073.2005
125. Tseng CT, Huang C, Newman P, Wang N, Narayanan K, Watts DM, Makino S, Packard MM, Zaki SR, Chan TS, Peters CJ (2007) Severe acute respiratory syndrome coronavirus infection of mice transgenic for the human Angiotensin-converting enzyme 2 virus receptor. J Virol 81(3):1162–1173. doi:10.1128/JVI.01702-06

Bacteriophages as Potential Treatment Option for Antibiotic Resistant Bacteria

Robert Bragg, Wouter van der Westhuizen, Ji-Yun Lee, Elke Coetsee and Charlotte Boucher

Abstract The world is facing an ever-increasing problem with antibiotic resistant bacteria and we are rapidly heading for a post-antibiotic era. There is an urgent need to investigate alterative treatment options while there are still a few antibiotics left. Bacteriophages are viruses that specifically target bacteria. Before the development of antibiotics, some efforts were made to use bacteriophages as a treatment option, but most of this research stopped soon after the discovery of antibiotics. There are two different replication options which bacteriophages employ. These are the lytic and lysogenic life cycles. Both these life cycles have potential as treatment options. There are various advantages and disadvantages to the use of bacteriophages as treatment options. The main advantage is the specificity of bacteriophages and treatments can be designed to specifically target pathogenic bacteria while not negatively affecting the normal microbiota. There are various advantages to this. However, the high level of specificity also creates potential problems, the main being the requirement of highly specific diagnostic procedures. Another potential problem with phage therapy includes the development of immunity and limitations with the registration of phage therapy options. The latter is driving research toward the expression of phage genes which break the bacterial cell wall, which could then be used as a treatment option. Various aspects of phage therapy have been investigated in studies undertaken by our research group. We have investigated specificity of phages to various avian pathogenic *E. coli* isolates. Furthermore, the exciting NanoSAM technology has been employed to investigate bacteriophage replication and aspects of this will be discussed.

Keywords Bacteriophage · Therapy · Antibiotic resistance · *Escherichia coli* · NanoSAM

R. Bragg (✉) · W. van der Westhuizen · J.-Y. Lee · E. Coetsee · C. Boucher
Department of Microbial, Biochemical and Food Biotechnology, University of the Free State, P.O. Box 399, Bloemfontein 9300, South Africa
e-mail: Braggrr@ufs.ac.za

R. Adhikari and S. Thapa (eds.), *Infectious Diseases and Nanomedicine I*, Advances in Experimental Medicine and Biology 807,
DOI: 10.1007/978-81-322-1777-0_7, © Springer India 2014

Increasing Problems with Antibiotic Resistance

The world is facing an ever-increasing problem with antibiotic resistance. This was one of the main underlying themes of the First International Conference on Infectious Diseases and Nanomedicine, which was held in Nepal in December 2012 [1]. The impact of this increased antibiotic resistance and the ban of the use of antibiotics in animal production have been covered in this issue [2] and also in other chapters in this issue.

In 1947, researchers first recognized antibiotic resistant bacteria when penicillin resistance was found in *Staphylococcus aureus*. By the 1960s, this antibiotic resistance had become widespread and clinically significant [3]. In 1996, some bacteria were reported to show resistance against third generation antibiotics such as cephalosporin and vancomycin [4]. The multiple antibiotic resistant bacteria of the 1990s are a possible sign of an impending post-antibiotic era, which poses a threat to human and animal health in the future.

As we are moving toward a post-antibiotic era, alternative methods for the control of bacterial diseases need to be investigated. There are a few potential alternatives which need to be more fully investigated. Improvement of biosecurity is one such option and this has been covered elsewhere in this issue [2].

Another option is the use of bacteriophages, which is the main subject of this chapter.

An ideal therapy to cure bacterial diseases must be highly effective while having no toxic effects to the host. Bacteriophages have shown no toxic effects in hosts, except for some rare, reversible allergic reactions and high success rates have been reported [4].

Review on Bacteriophages

Bacteriophages are the most abundantly available biomass on earth outnumbering prokaryotes by approximately tenfold in number [5]. They are viruses that recognize and target bacterial cells in order to replicate. The International Committee for Taxonomy of Viruses classifies viruses into three orders, 61 families, and 241 genera of which bacteriophages are included in the order Caudovirales consisting of 13 families and 30 genera [6]. Bacteriophages have capsid protein heads which carry and protect the genomic material of the viruses. The genomic material can vary in size, arrangement (circular, linear or segmented), and structure (ssDNA, dsDNA, ssRNA, dsRNA) depending on the virus [7]. Bacteriophages have been instrumental in the investigation and subsequent understanding of various molecular concepts.

The first reported discovery of bacteriophages was by Ernest Hankin who reported antibacterial activity against *Vibrio cholerae* in the Ganges and Jumna rivers, India, in 1896, which he described as a phenomenon caused by an unknown substance [8]. A British pathologist, Frederick William Twort, later observed bacteriophages in 1915 in London [6]. In 1917 a French Canadian, Fèlix Hubert d'Hèrelle, working in Paris, observed the lysis of *Shigella* cells in a broth culture. Twort did not pursue his discovery, but d'Hèrelle did. This led to d'Herelle naming the virus "bacteriophage," which is a combination of "bacteria" and "pagein," which means, "to eat" in Greek. d'Herelle later used the bacteriophages to treat dysentery which was the first reported therapeutic use of bacteriophages.

Bacteriophages are highly host specific and will only attack specific strains depending on the bacteriophage used and the target proteins which facilitate adhesion to the bacterial host. If the normal microbiota is not targeted, as in an ideal therapy, the normal microbiota will be able to compete successfully with other potential pathogens. This is not the case during antibiotic treatment during which normal microbiota and nonresistant pathogens are destroyed simultaneously.

Bacteriophages recognize their host through specific receptors on the bacterial host cell. After attachment, the genetic material of the bacteriophage is injected into the host. Once the phage genetic material is intracytosolic, it can either integrate into the host genome or remain as unintegrated extragenomic genetic material. The former life cycle is lysogenic while the latter is lytic. Virulent phages follow a lytic life cycle and take over the host's replicative mechanisms with the aim of successfully producing multiple copies of mature virus particles and ultimately lysing the host cell. Temperate phages revert to lysogeny and integrate into the host genome until they are induced into a lytic life cycle via mutagenic stress. They are also more host specific than are virulent ones as they are limited to the host in which they are integrated.

What has been revealed about the two potential life cycles of bacteriophages, as illustrated by our understanding of phage λ is the importance of the bistable genetic switch. At a glance, this switch comprises a repressor gene and Cro, *cI*, and *cro*, respectively, which are found next to each other [9, 10]. The presence of high amounts of the repressor protein CI allows the bacteriophage to remain inactive within the host cell, effectively resulting in a lysogen, i.e., the integration and maintenance of the bacteriophage nucleic acid within the host in the form of a prophage. In a detailed view of this switch, there are promoter regions and operator regions that are specific for the control of transcription for both *cI* and *cro* [9–11]. The promoter P_R aligns RNA polymerase to transcribe *cro*, while the P_{RM} promoter aligns it to transcribe *cI* [9]. The simplicity and elegance of this biswitch lies in the ability of both repressor and Cro to bind to the operator O_R which consists of O_R1, O_R2, and O_R3 [9]. This operator partly covers both the P_R and P_{RM} region. Upon the binding of repressor to O_R1 and O_R2, RNA polymerase binds and begins transcription at P_{RM} effectively transcribing *cI*. Upon a high concentration of repressor being present in the cell, it will eventually bind to O_R3 which has the weakest binding affinity for it. This shuts down P_{RM} and RNA polymerase is then inhibited from attaching to this region and transcribing *cI*, which is autoregulation that prevents the switch from producing unnecessary amounts of repressor. The level of repressor that is present within a lysogen is usually constant which effectively keeps the prophage integrated.

However, if the host DNA is damaged in the presence of a stimulating inducer such as ultraviolet irradiation, RecA, a bacterial protein, begins to cleave the repressor protein [9, 10]. RecA's function in the cell is to repair and maintain DNA and it is biochemically linked to repressor LexA which represses the bacterial SOS response mechanism [9, 12]. Once DNA is damaged and in need of repair, RecA assists in the cleavage of LexA which allows for the SOS response to begin and its genes to be transcribed. Since LexA and the λ repressor CI are structurally similar to one another, RecA also cleaves the latter when DNA is damaged [9–12]. This action releases the repressor from O_R and allows RNA polymerase to transcribe *cro* starting from P_R. Cro is converse to the repressor in that its highest binding affinity is for O_R3 which results in the repression of CI transcription. Thereafter, it will bind to O_R2 and then O_R1 with the accumulation of Cro at high concentrations. This too is an autoregulating negative feedback mechanism of the cell in preventing the unnecessary and excessive production of Cro. A study indicated that Cro strongly represses the P_{RM} and in this way converts the lysogenic state into a lytic one [13]. The function that Cro has in the cell is to ensure that this move from lysogeny to lysis transits efficiently and it is not necessary for the lytic state itself.

The bacteriophage–bacterial host liaison is an evolutionary relationship where resistance against bacteriophage attack is possible. In the same way that bacteria develop resistance against antibiotics, they also develop resistance to bacteriophages. Some methods that the host cell may employ include restriction digestion of the foreign phage DNA upon injection into the cell. The use of the CRISPR (clustered regularly interspersed palindromic repeats) via complementary binding to the phage DNA or subsequent RNA strands, effectively limiting replication. Another method is abortive infection that essentially destroys the infected host cell before bacteriophage virions are produced and in this way arrests the bacteriophage infection from spreading to surrounding healthy cells [14]. However, it has been shown that the bacterial host cell is able to revert to a susceptible state after a number of passages [15] which emphasizes the nature of bacteriophages as a sustainable tool to be used against bacterial infections.

Review on Phage Therapy

With the discovery and development of antibiotics, research on bacteriophage therapy declined, although in 1970, the Society of Friends of Felix d'Herelle, was founded to continue bacteriophage research [3]. In 1980, Dr. Stefan Slopek and his colleagues in Wroclaw, Poland, started clinical trials to treat patients infected with antibiotic resistant bacteria with bacteriophages. Other countries, like the former Soviet Union, continued research on bacteriophages and continued using them to treat serious bacterial infections which were resistant to antibiotics [3]. However, Western researchers continued with their research on antibiotic development after the discovery of antibiotics in 1928 [8].

One of the main disadvantages of the use of antibiotics is the ever-increasing problem with antibiotic resistance. One of the main advantages of the use of bacteriophages for the control of bacterial diseases is that the problems with resistance to bacteriophages are far less of a concern than the resistance to antibiotics. There are various reasons for this. First, it has been hypothesized that bacteriophages naturally evolved alongside their host bacteria [16]. Therefore, if a bacterium develops resistance to a bacteriophage, the bacteriophages could adapt through mutation and evolve to once again infect the resistant bacterium. Second, there are countless varieties of bacteriophages everywhere in the environment. If a particular bacterium develops resistance to one bacteriophage, there is a strong possibility that other bacteriophages could still infect that bacterium. With antibiotics, the number of different antibiotics is limited and it is not a trivial matter to find a new antibiotic if a bacterium develops resistance to the antibiotics.

Another advantage of bacteriophage therapy is that it has been demonstrated that when administered intravenously, bacteriophages can be found in nearly all organs [17] which is ideal when treating localized infections in different parts of the body. This is an advantage when compared to some antibiotics, which only reach certain and specific organs during treatment [17]. Injection of bacteriophages is not a suitable option in the poultry industry, due to the large number of birds involved, but it demonstrates that if the bacteriophages are capable of reaching the blood stream, they should reach all the organs.

Another major advantage of bacteriophage therapy is that bacteriophages quickly reproduce during the lytic life cycle, growing exponentially in number [18]. On average, bacteriophages can release approximately 100 new bacteriophages per lytic infection cycle which takes about 25 min in the bacteriophage T4's life cycle [19]. If each of the bacteriophages successfully infects a bacterium, there will be 0.1 billion bacteriophages after the fourth cycle. This allows for highly effective therapy, even if the administered dosage is low. This is a great advantage over antibiotic therapies since phage therapy is self-replicating, unlike the fixed dose of antibiotics used in treatment. Antibiotics can also cause fatal allergic reactions and can have severe side effects. Bacteriophages are nontoxic and have low occurrences of fully reversible allergic reactions, making therapy with bacteriophages very safe [3].

Bacteriophage treatments have the potential to be highly effective and an ideal therapy for bacterial infection, but there are drawbacks which need to be accounted for during the development and testing of the therapy.

Investigations into the Use of Phage Therapy for the Poultry Industry

Our research group has isolated various phages with varying specificity with the hope of selecting bacteriophages which are specific against avian pathogenic *E. coli* (APEC). The specificity of the isolated phages has been evaluated against a collection of avian pathogenic and nonpathogenic *E. coli* strains. The results of these specificity tests can be seen in Table 1.

Table 1 Bacteriophages' (*Row 2*) specificity profiles against various *E. coli* strains (Column 1). An "x" indicates the ability of the phage to lyse the appropriate strain of *E. coli*. The *last row* shows the total number of *E. coli* strains each bacteriophage is able to lyse. The *final column* shows the number of bacteriophages which are able to lyse each *E. coli* strain

Strains	Phages										Vir. Gene Total/18	Amount of phages lysing *E. coli* strain
	K12p	1080p	B98p	B98(3)p	1323p	1323 (3)p	B771p	B771(3)p	Zim11.6p	25922p		
WCD1		x		x	x					x	12	4
WCD2		x			x						12	2
WCD3		x		x	x					x	7	4
K12	x	x	X			x	x	x		x	5	7
O18	x	x	X	x	x	x	x	x		x	6	9
87							x	x			3	2
O142 /O125							x	x			5	2
O157:H7		x			x					x	6	3
172		x					x	x			5	3
214		x		x	x		x	x		x	5	6
25922		x			x		x	x		x	11	5
cos[f]								x			16	1
ExEgg											5	0
104	x	x	X	x	x	x	x	x		x	5	9
B98			X	x	x		x	x			12	5
31P	x	x	X		x	x	x	x			16	7
76					x		x	x			7	3

(continued)

Table 1 (continued)

Strains	Phages										Vir. Gene Total/18	Amount of phages lysing *E. coli* strain
	K12p	1080p	B98p	B98(3)p	1323p	1323 (3)p	B771p	B771(3)p	Zim11.6p	25922p		
B771							x	x			13	2
B841	x	x			x		x	x			16	5
1080											15	0
1304					x	x	x	x			17	4
1323		x			x	x	x				17	4
B1634							x				15	1
Zim11.6					x				x		6	2
Amount of *E.coli* strains lysed by phage	**5**	**13**	**5**	**6**	**15**	**6**	**16**	**15**	**1**	**8**		

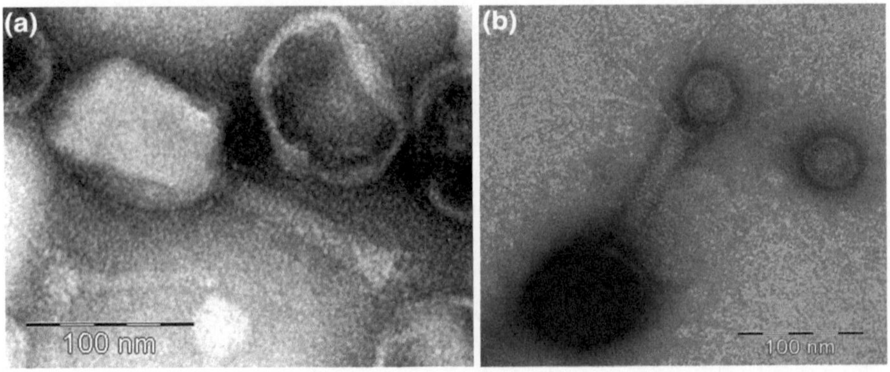

Fig. 1 Two TEM micrographs (**a**, **b**) of a T-even bacteriophage, designated B771p, negatively stained on a carbon-coated formvar-layered copper grid

Fig. 2 TEM micrograph of a section of sample embedded in epoxy (captured at 130 000x magnification) shows P4-like bacteriophages adhered to an *E. coli* cell on a formvar-layered copper grid

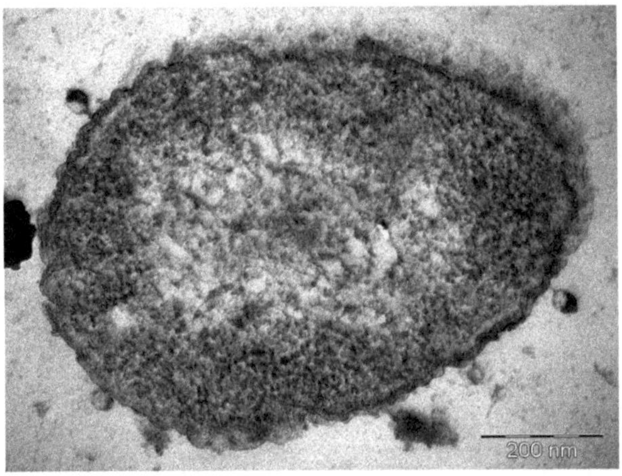

Transmission Electron Microscopy (TEM) was performed alongside molecular methods to classify the isolated bacteriophages (Figs. 1, 2), however this is an on-going process as more bacteriophages are isolated and added to our bacteriophage-library.

Bacteriophage screening was performed on APEC strains which virulence profiles have been determined [20]. From this information, bacteriophages can be selected which specificity profiles show higher specificity to APEC strains and more importantly, determining which bacteriophages can be used to lyse all known APEC strains in our collection. The future work will include challenging chickens with APEC and administering bacteriophages with the aim of treating and preventing infections.

As mentioned previously, our group has made use of NanoSAM technology on bacteriophages. NanoSAM technology has been reviewed in this issue [21].

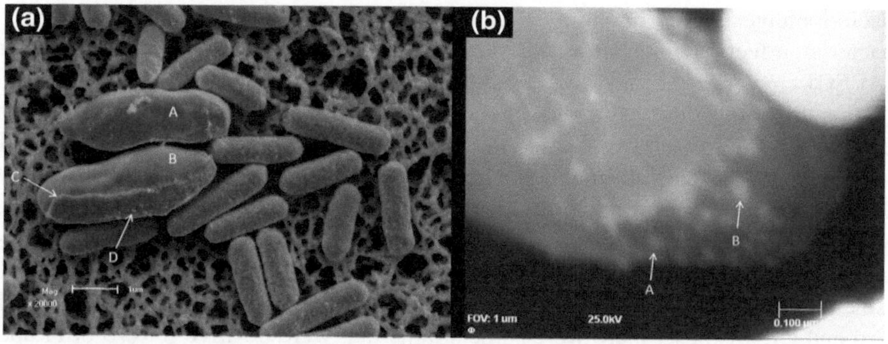

Fig. 3 SEM micrographs of *E. coli* cells which have been exposed to bacteriophage. **a** Enlarged cells were observed (cells *A* and *B*) and cell wall tearing is indicated by arrow *C*. Arrow *D* shows a protuberance suspected to be a bacteriophage. **b** An enlarged *E. coli* cell as in *A* which was subjected to Argon etching in the NanoSAM. Arrows *A* and *B* show protuberances that are suspected bacteriophages inside the cell, which have been exposed through the etching process

The NanoSAM was used in an investigation of the replication cycle of a P4-like bacteriophage which leads to the enlargement of *E. coli* cells [22]. In the study, scanning electron microscopy was used to observe these enlarged cells, and the Argon-ion gun of the NanoSAM was used to etch the surface of these enlarged cells to expose the bacteriophages present inside the cell (Fig. 3). Bacteriophage-like structures were found in these cells which prove to be much more resilient to etching in comparison to normal-sized cells.

Induction of Lysogenic Phages as Possible Candidates for Highly Specific Phage Therapy

Based on the relevance of phage therapy and the added benefit of bacteriophage–host specificity, part of our study involved detection and induction of prophages from lysogens. The interest behind this focus point was the aim of finding bacteriophages that are highly specific for a single host as opposed to a host range that consisted of several strains of a species. As the microbial community that lines the colon is made up of 30 genera and several hundred species besides [23], the more specific a treatment is for a particular pathogen the more effective it is. The microbial lining of the large intestine is important for health and contains beneficial and opportunistic microorganisms that lower the chance of pathogens from attaching and causing disease. Therefore, targeting specific pathogenic bacteria as a treatment while leaving the general microbiota unharmed would be useful. The most specific bacteriophages are the temperate ones which are integrated into the host genome as they are closely linked with the host for survival. As opposed to highly virulent bacteriophages, the temperate bacteriophages are proposed as being stringent for their host cells. Therefore, it was hypothesized that temperate bacteriophages would be the ideal candidate for highly host specific treatment of bacterial infection with an aim of targeting Avian Pathogenic *Escherichia coli* (APEC).

For this study, the *cro* and *cI* genes of λ phage were identified and primers designed on these genes were used in PCR to target and screen for the presence of these genes. The reason for this is because they encode for Cro and CI that play very important roles in the ability of the prophage to enter into a lytic life cycle. Without the presence of these, especially Cro, it is highly unlikely that the prophage will be able to become lytic. From the *E. coli* collection at our laboratory, 31 isolates containing both APECs and non-APECs were screened for the presence of Cro and CI. Of these, there were some strains that were positive for *cro* or *cI* and others that screened positive for both of the genes which was an indication that there were either lambdoid prophages or remnants of prophages present. Those that were identified as having both of the genes present were further used in induction studies in order to try and induce the bacteriophages into a lytic cycle. Methods of induction included shortwave UV, heat, and mitomycin C induction. All three induction methods were done over various ranges. For the UV induction study, a range of different exposure times was used. For the heat induction, a range of 40–60 °C was used and for the mitomycin C induction, a combination range of pH and mitomycin C concentration was used. Using the agar overlay method, there were some plaques that were visible after induction using UV and mitomycin C, but not with heat induction. However, the plaques could only be processed up to a maximum of three passages before they were not visible anymore. Furthermore, induction of the strains of *E. coli* using the same induction parameters often did not yield bacteriophage plaques again. This made the progress of the study difficult as the reproducibility of the experiments was unreliable. Although further work into the project will continue, we are currently approaching the viability of phage therapy from another angle with the investigation into and production of bacteriophage endolysins.

Identification of Bacteriophages in *Avibacterium paragallinarum*

Avibacterium paragallinarum is a bacterium which causes infectious coryza in poultry. This disease results in a drop in egg production which can be as high as 40 %. This bacterium can have a very serious economic impact in the layer industry.

Vaccine development against this disease is dependent on the correct identification of the different serovars of the bacterium as there is no cross protection between the different serovars [24].

There are limitations on the use of antibiotics for control of this disease in laying birds, as once the birds are placed onto antibiotics, the eggs should not be sold for human consumption until the end of the antibiotic withdrawal periods which are prescribed by most of the antibiotics currently available for use in layer chickens.

The use of bacteriophages as a potential treatment option for infectious coryza holds much promise. Prophage sequences have been found in other members of the *Pasteurellaceae* family. Therefore, it is likely that there might be prophages present in the different *A. paragallinarum* serotypes. The family *Pasteurellaceae* includes the *Haemophilus, Actinobacillus, Pasteurella, and Mannheimia* genera of bacteria, which causes a variety of diseases in humans and animals [25]. At least two prophages have been found in *Mannheimia haemolytica.* Both these prophages encode several Mu orthologs [26]. Studies conducted by Froshauer et al. [27] indicate that the antibiotic danofloxacin could induce a prophage in a serotype A1 isolate of *M. haemolytica*.

There are numerous other reports of bacteriophages present in the *Pasteurellaceae* family [28–32]. These prophages include the following; the genomes of two phages HP1 and HP2 isolated from *H. influenzae* have been sequenced and both are members of the P2 family of temperate bacteriophages [28]. Also the complete genome of a lambdoid temperate bacteriophage was reported to be found in *Actinobacillus actinomycetemcomitans* [29]. There has also been a Mu-like prophage identified within the genome of the *H. influenzae* Rd strain [30] and HP1-like sequences were also reported in *H. somnus* [31].

It has been shown that the presence of prophages has an effect on the virulence and pathogenicity of bacterial species. One such example is in the case of avirulent strains of *Corynebacterium diphtheriae* that is infected with a bacteriophage yielded virulent lysogens that produced the diphtheria toxin which causes diphtheria in humans [33]. Another example of this is the production of the scarlatina exotoxin by a temperate bacteriophage within the genome of non-toxigenic streptococci [33, 34]. Therefore, the presence of prophages within the genome of *A. paragallinarum* might explain the difference virulence levels of the different serovars. According to work done by Roodt and co-workers [32], two complete prophages have been assembled for *A. paragallinarum* C-2 serovar. One of these prophages resembles a Mu-phage (ΦAvpmuC-2M) and the other a HP2 phage (ΦAvpC-2M-HP2) that are present in *H. influenzae*. It would be interesting to determine whether there are similar prophages present in the other serovars of *A. paragallinarum*.

The findings of these prophages in *A. paragallinarum* might also contribute to the understanding or explain the occurrence of different serotypes in this bacterium. This might also shed light on the existence of NAD^{+}independent *A. paragallinarum* strains. Both bacteriophages and plasmids have the capability to enhance the pathogenicity of microorganisms [35]. Thus, the appropriate mechanisms of the emergence of new and more virulent serovars might be provided by presence of these elements [35]. The presence of these phages can also serve as an alternative treatment option against IC in chickens.

Conclusions

The use of bacteriophages for the treatment of bacterial diseases in a post-antibiotic era holds much promise. In order to fully evaluate the potential use of bacteriophages as a treatment option, it is essential that all aspects of the biology of the phages be fully understood. Our research group has been focusing on the use of bacteriophages for the treatment of bacterial diseases in the poultry industry. Specificity of lytic phages against *E. coli* has been studied. Efforts have also been made to induce lysogenic bacteriophages from *E. coli*. This approach met with some success, but it was found that the phages reverted to a lysogenic state after only a few passages. Lysogenic phages have also been identified in *A. paragallinarum* and attempts will be made to induce these phages.

As a result of the very level of specificity identified among the lytic phages, and the problems encountered with the stable induction of lysogenic phages, the emphasis of our research is now moving toward the expression of selected genes from bacteriophages in the hope that these expressed antigens can lyse bacterial cells.

References

1. Bragg RR, Kock L (2013) Nanomedicine and infectious diseases. Ex Rev Anti-infect Ther 11:359–361
2. Bragg RR, Jansen A, Coetzee M, van der Westhuizen W, Boucher CE (2013) Bacterial resistance to quaternary ammonium compounds (QAC) disinfectants. Advs. Exp Medicine, Biology: Adhikari and Thapa (Eds) Infectious Diseases and Nanomedicine I Chapter 7
3. Alisky J, Iczkowski K, Rapoport A, Troitsky N (1998) Bacteriophages show promise as antimicrobial agents. J Infect 36:5–15
4. Tenover FC, Hughes JM (1996) The challenges of emerging infectious diseases: development and spread of multiply-resistant bacterial pathogens. J Am Med Assoc 275:300–304
5. Brüssow H, Hendrix RW (2002) Phage genomics: small is beautiful (Minireview). Cell 108:13–16
6. Ackermann HW (2003) Bacteriophage observations and evolution. Res Microbiol 154:245–251
7. Ackermann HW (2007) 5500 phages examined in the electron microscope. Arch Virol 152:227–243
8. Sulakvelidze A, Alavidze Z, Morris JG Jr (2001) Bacteriophage therapy (Minireview). Antimicrob Agents Chemother 45:649–659
9. Ptashne M (1992) A genetic switch. 2nd edn. Cell Press, Cambridge, pp 13–30
10. Dodd IB, Shearwin KE, Barry EJ (2005) Revisited gene regulation in bacteriophage lambda. Curr Opin Genet Dev 15:145–152
11. Arkin A, Ross J, McAdams HH (1998) Stochastic kinetic analysis of developmental pathway bifurcation in phage λ-infected *Escherichia coli* cells. Genetics 149:1633–1648
12. Witkin E (1976) Ultraviolet mutagenesis and inducible DNA repair in *Escherichia coli*. Bacteriol Rev 40:869–907
13. Schubert RA, Dodd IB, Egan JB, Shearwin KE (2007) Cro's role in the CI-Cro bistable switch is critical for l's transition from lysogeny to lytic development. Genes Dev 21:2461–2472

14. Stern A, Sorek R (2011) The phage-host arms-race: shaping the evolution of microbes. BioEssays 33:43–51
15. O'Flynn G, Ross RP, Fitzgerald GF, Coffey A (2004) Evaluation of a cocktail of three bacteriophages for biocontrol of *Escherichia coli* O157:H7. Appl Environ Microbiol 70:3417–3424
16. Hendrix RW, Smith MC, Burns RN, Ford ME, Hatfull GF (1999) Evolutionary relationships among diverse bacteriophages and prophages: all the world's a phage. Proc Nat Acad Sci 96:2192–2197
17. Dabrowska K, Switala-Jelen K, Opolski A, Weber-Dabrowska B, Gorski A (2005) Bacteriophage penetration in vertebrates. J Appl Microbiol 98:7–13
18. Carlton RM (1999) Phage therapy: past history and future prospects. Archi Immunol Ther Exp 47:267–274
19. Madigan MT, Martinko JM (2006) Essentials of virology. In: Carlson J (ed) Brock biology of microorganisms, Pearson Prentice Hall, Upper Saddle River, pp 244–245
20. Van der Westhuizen WA, Bragg RR (2012) Multiplex polymerase chain reaction for screening avian pathogenic *Escherichia coli* for virulence genes. Avian Pathol 41:33–40
21. Swart CW, Pohl CH, Kock JLF (2013) Auger-Architectomics: introducing a new nanotechnology to infectious disease. Advs. Exp Medicine Biology: Infectious Diseases and Nanomedicine. Chapter 1
22. van der Westhuizen WA, Kock JLF, Coetsee E, van Wyk PWJ, Swart HC, Bragg RR (2013) Investigation of the final stages of a P4-like coliphage infection in *Escherichia coli* through scanning, transmission and nano-scanning-auger electron microscopy (NanoSAM). Sci Res Essays 8:382–387
23. Canny GO, McCormick BA (2008) Bacteria in the intestine, helpful residents or enemies from within? Infect Immun 76:3360–3373
24. Bragg RR, Coetzee L, Verschoor JA (1996) Changes in the incidence of the different serovars of *Haemophilus paragallinarum* in South Africa: a possible explanation for vaccination failures. Onderstepoort J Vet Res 63:217–226
25. Highlander SK, Weissenberger S, Alvarez LE, Weinstock GM, Berget PB (2006) Complete nucleotide sequence of a P2 family lysogenic bactriophage, ΦMhaA1-PHL101, from *Mannheim haemolytica* serotype A1. Virology 350:79–89
26. Gioia J, Qin X, Jiang H, Clinkenbeard K Lo, Reggie Liu, Y Fox GE, Yerrapragada S, McLeod MP, McNeil TZ, Hemphill L, Sodergren E, Wang Q, Muzny DM, Homsi FJ, Weinstock GM, Highlander SK (2006) The genome sequence of *Mannheima haemolytica* A1: insights into virulence, natural competence and *Pasteurellaceae* phylogeny. J Bacteriol 188:7257–7266
27. Froshauer A, Silvia AM, Chidambaram M, Sharma B, Weinstock GM (1996) Sensitization of bacteria to danofloxacin by temperate prophages. Antimicrob Agents Chemother 40:1561–1563
28. Williams BJ, Golomb M, Phillips T, Brownlee J, Olson MV, Smih AL (2002) Bacteriophage HP of *Haemophilus influenza*. J Bacteriol 184:6893–6905
29. Resch G, Kulik EM, Dietrich FS, Meyer J (2004) Complete genomic nucleotide sequence of the temperate bacteriophage AaΦ23 of *Actinobacillus actinomycetemcomitans*. J Bacteriol 186:5523–5528
30. Morgan GJ, Hatfull GF, Casjens S, Hendrix RW (2002) Bacteriophage Mu genome sequence: analysis and comparison with Mu-like prophages in *Haemophilus, Neiserria* and *Deinococcus*. J Mol Biol 317:337–359
31. Pontarollo RA, Rioux CR, Potter AA (1997) Cloning and characterization of bacteriophage like DNA from *Haemophilus sommnus* homologous to phages P2 and HP1. J Bacteriol 179:1872–1879
32. Roodt Y, Bragg RR, Albertyn J (2012) Identification of prophages and prophage remnants within the genome of *Avibacterium paragallinarum* bacterium. Sequencing 2012:1–5. Article ID 953609, doi:10.1155/2012/

33. Tinsley CR, Bille E, Nassif X (2006) Bacteriophages and pathogenicity: more than providing a toxin. Microbes Infect 8:1365–1371
34. Wagner PL, Waldor MK (2002) Bacteriophage control of bacterial virulence. Infect Immun 70:3985–3993
35. Prescott LM, Harley JP, Klein DA (2002) Microbiology 5th edn. McGraw Hill, Boston, pp 294–820

Editors Profile

Dr. Rameshwar Adhikari is Associate Professor at Central Department of Chemistry, Tribhuvan University, Kathmandu. He has been Invited Professor in Rouen University (France), Mahatma Gandhi University (India), and Tokyo Institute of Technology (Japan). He is member of Nepal Chemical Society and German Engineers Society. He has been bestowed several awards including Ambassador Scientist of Alexander von Humboldt Foundation, Fellowship of International Union of Pure, and Applied Chemistry and POLYCHAR World Forum International Materials Science Prize. He is President of Nepal Polymer Institute and has convened several International Congresses, Symposia, and Workshops. His research areas include nanomaterials, biopolymers, and biomedical materials. He has authored over 100 research papers in peer-reviewed journals, co-authored four textbooks, and edited two books.

Mr. Santosh Thapa completed his M.Sc. (Medical Microbiology) from Tribhuvan University, Kathmandu and taught related courses to undergraduate students at departments of microbiology of Tribhuvan University and affiliated colleges. He is the founder General Secretary of Nepalese Association of Medical Microbiology, members of American Society for Microbiology and Society for Applied Microbiology, and associated with International Union of Microbiological Societies. He has worked toward organizing several international conferences and workshops and has authored five research papers in peer-reviewed journals. Currently, he is at Graduate School of Biomedical Sciences, University of North Texas Health Science Center, Fort Worth, Texas, USA for his Ph.D. study.

R. Adhikari and S. Thapa (eds.), *Infectious Diseases and Nanomedicine I*,
Advances in Experimental Medicine and Biology 807,
DOI: 10.1007/978-81-322-1777-0,

Authors' Profile

Dr. Kaushik [illegible] Adhikari is Associate Professor at Central Department of Chemistry, Tribhuvan University, Kathmandu. He has been trained from several [illegible] University [illegible] [illegible] [illegible] [illegible] and [illegible] Institute of Technology [illegible] 100 [illegible] [illegible] [illegible] [illegible] [illegible] [illegible] [illegible] [illegible] [illegible] [illegible] [illegible]

[illegible]

Author Index

R. Adhikari and S. Thapa (eds.), *Infectious Diseases and Nanomedicine I*,
Advances in Experimental Medicine and Biology 807,
DOI: 10.1007/978-81-322-1777-0, © Springer India 2014

Author Index

Subject Index

R. Adhikari and S. Thapa (eds.), *Infectious Diseases and Nanomedicine I*, Advances in Experimental Medicine and Biology 807,
DOI: 10.1007/978-81-322-1777-0, © Springer India 2014